A HORSE'S *LIFE*

THE NEUROSCIENCE OF EQUINE WELFARE

DR. STEPHEN PETERS

MARK RASHID

CRISSI MCDONALD

Paperback ISBN: 979-8-9953285-0-6
Hardcover ISBN: 979-8-9953285-1-3
Ebook ISBN: 979-8-9953285-2-0

Library of Congress Control Number: 2026907416

Published by Elk Tooth Books
1820 Graham Ln, Estes Park, CO 80517
author.ahorseslife@gmail.com

Book Build by WestSky Studio
Broken Arrow, Oklahoma
www.westsky.studio

Printed in the United States of America

First Edition, 2026

Elk Tooth Books

PRAISE FOR
A HORSE'S LIFE

"Grounded in moving stories and relationships with real horses, this book is the best of all worlds for students of the horse. While the authors' decades of advanced expertise are clear, they present genuine heartfelt explorations of scenarios we can all resonate with, threaded with respect and compassion for the horses. With perfectly complementary elucidations of horse physiology, nervous systems, and behavior science expressed within each story, the result is a guided invitation for readers to connect with the mental, physical, emotional, relational unity of horses. Reading this book, I feel grounded, educated and inspired."

–Dr. Jessica White Plume, *Oglala Lakota*

"I almost couldn't come up with the words to express not only how important this book is, but how enjoyable and effortless it made understanding the science behind how your horse experiences the world – and your part in it. The experiences of these three knowledgeable and talented writers and the science behind it made complete sense. I was learning horse brain science without straining."

–Jim Masterson, *Founder of The Masterson Method®*

"In *A Horse's Life,* the authors bring together neuroscience and horsemanship in a way that is both rigorous and deeply accessible. Through relatable, real-world narratives and clear scientific explanation, they illuminate how horses perceive, process, and respond to their environment and to humans. By blending practical experience with contemporary science, the book translates what has long been called "common horse sense" into a framework grounded in expertise, brain science and species-specific behavior. Thoughtfully written and immediately applicable, *A Horse's Life* equips owners, trainers, veterinarians, and equine professionals with knowledge to make well-informed and ethical decisions supporting equine well-being. This work bridges science and art, offering clarity and credibility to all who are committed to understanding horses. Simply put...It belongs on every equestrian's bookshelf."

—Jillian Kreinbring, *Equine Functional Anatomist and Clinician*

"This is the book I have been waiting for! Mark, Crissi and Dr. Peters are more than the sum of their parts. Stories and science are combined in a way that is relatable and readable. The lessons that inspire me most are ones that teach me to look at things from a new angle, and ones that remind me to slow down. This book does both. Educational and inspiring. One of the best horse books ever written!"

—Tik Maynard, *Canadian Three-Day Event Rider, Clinician , Two-time winner of Road To The Horse, Author of "Starting In The Middle," and "In The Middle Are The Horsemen."*

"A Horse's Life captures something rare—the ability to tell a beautiful story while quietly teaching important truths. As you read, you'll recognize yourself in the examples and walk away with insights that genuinely support where you are right now. Imagine the storytelling of Mark Rashid, the thoughtful approach of Crissi McDonald, and the scientific understanding of Dr. Stephen Peters, brought together in one place. That's what makes this book such a meaningful and memorable read. I hope you enjoy it as much as I have!"

—Josh Nichol, *Clinician and Educator*

"An understanding of equine brain science is the biggest innovation for horses in generations. Each chapter begins with a case study, written with clarity and compassion by Mark or Crissi. Then Dr. Peters follows with clinical notes, using brain science to explain what our horses can't tell us. A Horse's Life is important information for horse owners who continue to learn and do ever better for their horses, but also invaluable as a reference aid for equine professionals."

—Anna Blake, *Clinician and Author*

TABLE OF CONTENTS

INTRODUCTIONS

CRISSI MCDONALD

In the early 2000s, I searched for information about how horses' brains worked, horse behaviorism, or horse (for lack of a better word) psychology. I found two books, both of which used behavior and a very small bit of science to extrapolate information about how horses saw the world. While helpful, they didn't quite answer the many questions I had.

Years later, a friend of mine handed me a book by Dr. Stephen Peters and Martin Black called *Evidence-Based Horsemanship* and said, "You need to read this." She was right; here was a book that answered questions I didn't know I had. But even in 2017, I couldn't find any workshops with Dr. Peters, and he didn't have other books. I was hungry for more knowledge on the inner workings of horses. Hungry for information that was based on scientific evidence, as opposed to the most common answer I got to my questions, "We've always done it this way." I used the information in Dr. Peters' book as much as possible, while also continuing my commitment to being a student of the horse.

2018 rolls around, and one day, Mark came into our living room and said he'd received an email from a Dr. Stephen Peters, complimenting him on challenging the notion that horses understand the abstract concept of respect. I was instantly thrown back into the past, to the horsewoman I was in 2009, searching for information on horse brains.

After a series of emails between Mark and Steve, he invited us to dinner at his house in Mancos, the little town in Colorado we were working in that weekend. Mark and I and our friend Gray found ourselves sitting at Steve's kitchen table, completely speechless (and well fed) by the three-hour horse brain presentation he served with dinner.

The next day was the last day of a clinic, and we usually make a point to get to sleep early. After Steve's presentation, however, we didn't make it back to our trailer until almost midnight. I was buzzing with so much new information that I could barely sleep.

Here we are almost ten years later, and our friendship has evolved into collaborative clinics and, now, this book. My intention is that our book will

help improve your relationship with your horse through growing your understanding of who they are as themselves, instead of the stories we so easily create about them.

Horses give us their lives because they don't have any choice. If we're fortunate, they become willing partners. Repaying their service to humans by understanding them as much as we can is not, perhaps, a fair trade but certainly a worthwhile one.

MARK RASHID

Back in May of 2018, I received an email from Dr. Stephen Peters. It was a bit of a surprise only because just a few weeks earlier I had finished reading *Evidence Based Horsemanship* by Dr. Peters and horseman Martin Black. Like many people, I found it very interesting and informative.

In his email, Dr. Peters mentioned he had read my book *Finding the Missed Path: The Art of Restarting Horses.* I had written a chapter in the book challenging the idea that horses understand the concepts of respect and disrespect when it comes to their interactions with humans.

My thoughts on the subject were initiated from an inherent discomfort I had that many horse people had blindly accepted the idea that horses were capable of understanding the concepts of good and bad, right and wrong, respect and disrespect and then used it as a way to perhaps justify punishment for unwanted behavior in their horse. This discomfort was confirmed when doing research on the equine brain prior to writing the chapter.

What I had found was the horse doesn't possess the neural capability to understand the world the way humans do, and specifically, they are unable to understand human-made abstract concepts. As a result, humans are often punishing horses for things that the horse is incapable of understanding.

Steve expressed in his email that the information I passed along in that chapter was exactly how the horse's brain worked. That email began a conversation between us that ultimately culminated in him inviting Crissi and me, along with our senior assistant instructor, Gray Kyle-Graves to dinner at his house, which just happened to be near where we were doing a clinic.

Steve met us at the door and invited us in, and a few minutes later, we found ourselves in the middle of Steve's now world-famous, three-hour PowerPoint presentation on equine neuroscience. Following the presentation (and an amazing dinner), Steve and I had the opportunity to sit for another couple hours and visit about horses, horse behavior, and brain function.

The next morning at the clinic venue, I went about the business of taking care of our horses. On this morning, like pretty much all mornings, as I tossed hay over the fence, I did a cursory glance at the horses to make sure everybody was upright and nobody was bleeding from some overnight scuffle that might have occurred. Everything seemed to be in order, so I pushed the wheelbarrow into the pen and started to clean up.

I was thinking about everything I'd heard and seen the night before during Steve's talk and our ensuing conversation and how many connections had been made about horse behaviors I'd seen over the years and the correlation between that and what was happening from a neurologic standpoint. It was about then that something strange occurred.

I looked up at our horses, this time with more than the original cursory glance, and what I saw is hard to explain. While I was obviously still looking at a horse, it felt as if I was seeing an entirely different animal—more than just the flesh and bone, something deeper, more profound. It was as if I had previously seen horses in two-dimensional black and white, but suddenly they appeared in three-dimensional color.

A shift occurred for me that day that has permeated everything I do with horses from then till now. Instead of just trying to shape or reshape certain behaviors, now my first thought is, *How can I reframe this from a neurological standpoint?*

Today, Steve, Crissi, and I collaborate whenever time allows, and this book, *A Horse's Life: The Neuroscience of Equine Welfare*, is a byproduct of our shared time together. We hope you enjoy reading it as much as we've enjoyed putting it together and that perhaps it can help you see your horse in a little different light as well.

DR. STEPHEN PETERS

A Horse's Life is an examination of how horses are shaped by their lived experiences and how those experiences become wired into the nervous system. When those patterns support calm, clarity, and safety, the horse can learn, connect, and thrive. When experiences lead to confusion, stress, or fear, the nervous system adapts in ways that may appear as reactivity, shutdown, or ingrained defensive patterns. The good news is that the equine brain remains plastic throughout life. Patterns that no longer serve the horse can be rewired through appropriately timed experiences of safety, clarity, and relational stability. Through real stories, this book reveals the underlying neural mechanisms

that govern equine learning, emotional regulation, and behavioral expression. It offers a framework that prioritizes the horse's mental well-being and honors their natural learning processes.

My first encounter with Mark Rashid was through his book *Finding the Missed Path: The Art of Restarting Horses.* In it, Mark challenged a familiar belief in the horse world, that horses interpret our actions in terms of "respect" or "disrespect." He pointed out that horses are not making moral evaluations or conceptual judgments. They benefit from what has been made clear, consistent, and safe for them, and they avoid what has been confusing, uncomfortable, or overwhelming.

Mark's point was simple: horses do what works for them. If moving into our space has been reinforced, intentionally or not, they will continue to do it. If a boundary has been made clear and consistently maintained, they will stay within it. If the boundary has been unclear or inconsistently enforced, the horse has no reason to recognize it. This is not attitude, defiance, or a personality flaw. Horses do not think in these terms. They are responding to lived experience and nervous system state.

From a clinical neuroscience perspective, this is precisely how the brain works. Behavior is shaped through neuroplasticity—how neural circuits strengthen or weaken through repeated experience—not through abstract reasoning about human intention. When I finished Mark's book, I emailed him to share that his conclusions aligned directly with neuroscience.

Soon after, Mark taught a clinic near my home in Mancos, Colorado. I invited him, his wife, Crissi McDonald, and their senior instructor, Gray Graves, to dinner. I had planned a brief, informal presentation on equine neuroscience—how the horse's brain perceives, processes, and responds to the world around it. As we talked, it became clear that Mark's years of practical experience working with horses reflected the same processes we recognize in the neuroscience literature.

During this same period, Mark was collaborating with Jim Masterson. I had met Jim years earlier, when he approached me with questions about the neurological basis of the subtle responses seen in the Masterson Method. He was asking whether these responses were reflexive, conditioned, or part of a broader shift in autonomic balance. His questions were directly tied to the neurobiology of stress regulation and recovery.

As the four of us (Mark, Crissi, Jim, and I) continued our conversations, we realized we were all describing the same nervous system processes from different vantage points: horsemanship, bodywork, and neuroscience.

By bringing these perspectives together, we could offer a clearer and more complete understanding of the horse's experience.

This collaboration became the clinic *Your Horse's Brain: A User's Manual*, which we continue to teach annually at Happy Dog Ranch. The clinic includes equine brain dissections, current neuroscience, and live horse interactions in which Mark and I discuss what is unfolding in real time. Jim contributes his Masterson Method approach, and Crissi brings her lifelong experience with horses and the depth and clarity in how these stories and insights are presented and understood.

The idea for this book grew naturally from that work. Real horses with real histories show us how behavior develops over time, how learning is shaped by nervous system state, and how change can occur when conditions support it. Rather than writing a textbook or training manual, we chose to offer case-based narratives paired with Clinical Notes that explain the relevant neuroscience. This allows the science to be understood *in context*, where it matters most.

A central theme of this book is that learning and connection depend on the horse's ability to regulate their nervous system. When a horse perceives safety, neuroplasticity opens the door to new experiences and new learning. When the nervous system is overwhelmed, the brain shifts into protection, and defensive patterns take over. Understanding this is not merely academic; it is foundational to equine welfare.

WHY THIS MATTERS NOW

In human neurology, case studies have long been used to understand the brain. Oliver Sacks illuminated how neurological differences shaped perception, behavior, and identity. Similarly, the horses in this book are not conceptual examples; they are individuals whose lived experiences shaped their nervous systems, and thus their behavior. Their stories show how memories, associations, and physiological states become encoded and how change is possible.

Today, equine welfare is being reassessed in light of animal cognition and emotional experience. Welfare is not only physical, but also psychological. To work humanely with horses, we must understand how their nervous system forms patterns, why behaviors develop, and how learning occurs.

This understanding also shapes the public's trust—our social license to work with horses at all. As expectations evolve, humane and science-informed horsemanship is not only ethical, but also essential.

Our aim in this book is to provide a practical and compassionate framework for understanding the horse from the inside out. By beginning with the lived story of the horse and then examining the underlying neuroscience, we can deepen our understanding of how to support learning, connection, and well-being.

This is *A Horse's Life: The Neuroscience of Equine Welfare*. A view of the horse shaped by experience, understood through the nervous system, and supported through care rooted in science and empathy.

THE BRAIN BEHIND THE BEHAVIOR

DR. PETERS

Before we can fully understand a horse's behavior, emotional state, or capacity to learn, we have to understand the organ that makes those experiences possible. The equine brain is not built for obedience. It is built to keep a prey animal alive: to scan the environment, detect what matters, predict what might happen next, and guide action in ways that protect safety and preserve connection.

Every experience a horse has, be it peaceful or stressful, clear or confusing, passes through this system. Over time, experience doesn't just change behavior; it changes the brain. Patterns that reliably end in safety strengthen regulation and flexibility. Patterns that end in confusion, conflict, or fear strengthen protection and readiness. In that way, the horse's brain becomes a living record of its environment.

A useful way to understand the brain is in three broad regions that work together continuously: the hindbrain, the midbrain, and the forebrain. The hindbrain includes the brainstem, which governs essential life functions such as breathing, heart rate, blood pressure, digestion, and sleep–wake rhythms. It also includes the cerebellum, a major center for balance, timing, and the fine coordination of movement. These structures form the foundation; they keep the body regulated and ready.

Just above that foundation are fast, highly efficient systems for alertness and orientation, an advantage for an animal whose safety depends on noticing small changes quickly. Before a horse has time for deliberate evaluation, the body may already be preparing to orient, to mobilize, or to settle. What we call "reactivity" is often the nervous system doing exactly what it was built to do: detect significance and prepare the body to respond.

The forebrain supports more complex processing, including emotional evaluation, learning, memory, and decision-making. Within it, a network often referred to as the limbic system assigns emotional meaning to experience. The amygdala is especially important here. It monitors for significance, novelty, uncertainty, or potential threat, and it helps prioritize what the brain should remember and respond to quickly in the future. This is why a single frightening event can sometimes produce a lasting behavioral change: the nervous system learns that the cost of missing danger is too high.

At the same time, repetition builds efficiency. Another key system involved in this kind of learning is the basal ganglia. The basal ganglia are deep nuclei, not cortex. Most lie in the forebrain, but key parts of the circuit extend into the midbrain, including the substantia nigra. Functionally, these loops help select actions and store patterns so responses can become smoother, faster, and less effortful with practice. This is how horses develop coordinated movement, routines, and anticipatory patterns. Those habits can become calm and reliable when learning is clear and consistent, or defensive and rigid when repeated experience demands protection.

Flowing through all of these regions is the autonomic nervous system, which governs internal state. The sympathetic branch prepares the body for action by raising heart rate, increasing muscle tone, and sharpening sensory readiness. The parasympathetic branch supports recovery through rest, digestion, tissue repair, and social settling. In a well-supported nervous system, a horse moves fluidly between these states as the environment changes. In a chronically unpredictable or threatening environment, the system can stay stuck in readiness, and the horse may lose access to ease.

This brings us to the central insight: the brain is always updating. Every interaction, every approach, every release of pressure, and every moment of clarity or confusion delivers information to the nervous system. Repeated information becomes a pathway. Pathways become a pattern. Patterns become the horse's default expectations.

When a horse lives in conditions marked by consistency, predictability, and fair communication, the nervous system learns that the world is manageable. Curiosity rises. Learning becomes easier. Connection deepens. When con-

ditions are conflicting, overwhelming, or chronically uncertain, the nervous system adapts toward protection. The horse may look reactive, dull, resistant, or shut down, not because of attitude, but because its brain has been shaped to prioritize survival over engagement.

The stories in this book are case studies of living brains adapting, sometimes under pressure, sometimes in recovery, and sometimes in remarkable transformation. The changes you will read about begin inside the nervous system first. Behavior is the visible outcome; the brain is the origin.

And the good news is that the equine brain is not fixed. It is adaptive and responsive to new patterns of safety and clarity. With informed care and thoughtful handling, a horse's life can change because the brain that generates that life can change too. This is the science that underlies every story that follows.

THE HORSE'S NERVOUS SYSTEM:

Understanding What We're Seeing

TUFF

The Educated Horse

CRISSI

Tuff came to us as a seven-year-old from our friend Lloyd, a rancher and Quarter Horse breeder in Minnesota. Tuff was 14'3, a dun with a brown line down his back and zebra striping on his legs. He'd been used as a ranch horse and breeding stallion until he busted down some fences to get to mares and got the right side of his face kicked in for his trouble. After the vet told Lloyd that Tuff would be okay but would probably always have a dent in his face, Lloyd asked the vet to geld Tuff. It wasn't too long after this that Mark bought him from Lloyd.

Tuff's years as a ranch horse were apparent when Mark used him as a clinic horse; he could be around other horses without any worry, he had ponying and roping skills, and he could be hauled long miles without anxiety, as well as be ridden long miles on the trail without incident. Tuff worked cattle, crossed bridges and water, and would walk, trot, and canter quietly when asked. His lateral work was immediate and accurate; if asked to put one hoof slightly to the left, he would do that then wait for the next request.

When we hosted five-day clinics near our home, Tuff was one of the horses who would be assigned to a rider for the week. Over the years, Mark and I heard many different opinions about the kind of horse Tuff was. Most of the time, we nodded and smiled because most of the time our riders were saying things such as, "He sucks back behind my leg," "He's so slow," "I don't know where the 'go' button is," and other comments that arose (most likely) out of frustration. One comment we heard consistently was any number of variations of "Is he shut down?" Even in the mid- 2000s, this term was gaining traction.

Though I knew Tuff wasn't shut down, I could understand how folks could come to that conclusion. He stood at the hitch rail half asleep, one hind foot cocked. His ears were relaxed, his eyes half-closed. When being groomed and saddled and led into the arena, he didn't change much. Horses who were trailered in had no effect on him. Whether they were quiet, or pacing and whinnying, it was as if Tuff didn't register it. Loud noises, thunderstorms, horses cantering by him at speed, flapping tarps, out-of-balance riders; nothing bothered him. Some riders would call him "bombproof." Others would call him "sleepy" or "checked out."

Even having a new rider every week, he was always himself, never showing any confusion that the previous rider may have experienced as they learned. New horses would come and go on the property, and our herd, kept in a separate paddock, was quiet under his confident leadership.

Mark and I also heard comments about how fun Tuff was, how safe he felt, and how much they appreciated him showing them a deeper way to be with a horse. If Tuff had a rider who asked to do things with him, he was all in. If, however, he had a rider who wanted him to go while they sat on him, he wouldn't move. If he did, it was as if his hooves were pulling the earth with him.

We could put children on him, advanced riders, or beginners, never once worrying about their safety. It was as if upon meeting his rider, he was able to assess their skill level, and he adapted by the time the rider sat in the saddle and he had taken the first steps away from the mounting block.

I took him out on the road with us one year as we traveled to California, then on to Oregon and Washington. I'd watched other riders work with him for years and was eager to spend some time with him myself. Sure enough, as soon as I got on, I recognized that here was a horse who would require me to improve what I was offering.

At that point in my riding career, I'd ridden mostly forward horses, so I was constantly finding ways to quiet myself to keep their worry low. What Tuff taught me is that there is a vast difference between being quiet and being absent. He taught me how to be truly committed to what I wanted to work on with him, whether it was trotting, lateral work, cantering, or backing up. Toward the end of the six weeks we spent on the road together, we were doing trot and canter pirouettes, shoulder- and haunches-in, and walk-to-canter transitions, all with very little external aids, mostly with intention, presence, and breathing. There were many times when I felt Tuff's body and my own were reflections, one responding to the other without any barriers between us.

As he got older, we rode Tuff less. Mark and I began teaching clinics at Happy Dog Ranch, where they had other horses for folks to ride. We'd pull Tuff out several times a summer, and he would continue to teach me about softness and quiet presence.

There was a stretch of time that we didn't ride Tuff after he'd had an injury to his right front leg, so we treated him and let him be a horse. Two years went by, then we were asked to be Grand Marshalls for our town's local rodeo parade. I wanted to ride Tuff, who had recovered from his injury. We put shoes on him, and the next day, he and Rocky rode to the staging area for the parade. I know I was more nervous than Tuff (this was not long after my wreck where I'd sustained a traumatic brain injury).

Mark and Rocky and Tuff and I were in the middle of people clapping and cheering, firetrucks honking their horns and throwing candy. It was a sensory storm, surrounding us in explosive noises and flapping flags, a marching band, and other horses. I could feel Tuff's body relaxing underneath me, his breathing regular, the rhythmic sound of his hooves on the asphalt street. I began mimicking his breathing, admiring his brown-tipped ears that he'd flicked forward, his neck arched. He wasn't quite prancing, but the presence radiating from him was magnetic. I laughed as he glanced left and right, and I imagined him saying, "All this? For me?" It's one of my many favorite memories of Tuff.

As he got older, he became our go-to horse for the grandkids to ride. He was our herd's leader for many years, never using hooves or teeth when we brought new horses into the herd. With barely an ear flick or a tail swish, he was the example of quiet confidence that our other horses relied on, just as I had, for many years.

CLINICAL NOTES

DR. PETERS

Tuff exemplified a profile of high competence with low arousal, a state that is often mistaken for **"shutdown."** His quiet posture was not disengagement or defeat. It was the outward expression of an exquisitely regulated, well-educated nervous system. Rather than showing learned helplessness, a collapse of agency caused by prolonged, inescapable stress and marked by impaired learning, Tuff demonstrated the opposite: preserved agency, selective responsiveness, intact curiosity, and fully functional learning circuits.

This distinction matters because, from the outside, a shutdown horse and a horse in learned helplessness can look almost identical. Both may appear quiet, withdrawn, unmotivated, and emotionally flat. Both may stand with a lowered head, show minimal response to their environment, and display little initiative to move, explore, or engage. That similar outward appearance is exactly why appearance alone is not enough to make an accurate diagnosis. What differentiates transient shutdown from learned helplessness is not simply the behavior you see in the moment but the history and circumstances that shaped it.

A shutdown state typically follows temporary overwhelm, pain, fear, confusion, illness, or fatigue. In those cases, the nervous system has pulled back as a protective, energy-conserving measure, but the capacity for recovery and agency remains intact. When safety, comfort, and predictability return, the

SHUTDOWN

Shutdown is a word often used to describe a horse that appears quiet, dull, unresponsive, or withdrawn. It is a description of what we see on the outside, not a specific diagnosis of what is happening inside the brain. From a neurological perspective, shutdown is a temporary state in which the nervous system has reduced activity in order to conserve energy or protect itself. The horse may be tired, overwhelmed, dissociated, ill, or simply conserving resources. In these cases, the capacity for choice, learning, and motivation is still present, even if it isn't being expressed in the moment. When safety, rest, clarity, or comfort return, the horse can often re-engage.

horse begins to re-engage, often in small ways at first: brighter eyes, renewed curiosity, increased responsiveness, and an expanding interest in the world. Learned helplessness, by contrast, emerges only after prolonged, uncontrollable, inescapable stress. In that condition, the brain has learned that effort does not change the outcome. The sense of agency has been deeply eroded, and the lack of response often persists even when the environment improves and opportunities for choice are reintroduced. In short, behavior is only the final chapter; the true explanation lives in the pattern of experience over time.

In Tuff's case, the systems responsible for detecting threats and maintaining safety were operating efficiently and accurately. His amygdala–hippocampal circuits had been shaped by thousands of predictable, non-threatening experiences. Novel stimuli such as movement, sound, crowds, flapping objects, and sudden environmental shifts were evaluated quickly and classified as non-dangerous. His reticular activating system still registered incoming sensory information, but perceived significance remained low, so the orienting response softened rapidly rather than escalating into fear or hypervigilance. This is the signature of an accurately calibrated nervous system with refined predictive processing rather than exaggerated threat detection.

At the level of stress physiology, Tuff showed strong top-down modulation from frontal and cingulate networks in dialogue with thalamic and limbic systems. This organization supports a stable hypothalamic–pituitary–adrenal axis, and it is consistent with low, well-regulated cortisol output. Neurochemically, his internal state appeared optimized for calm clarity: serotonin supporting emotional steadiness, acetylcholine sustaining attention, and norepinephrine providing alertness without tipping into anxiety. This balanced internal landscape allowed subtle communication, precise movement, and thoughtful responsiveness even in stimulating environments.

Importantly, Tuff was not unmotivated. He demonstrated a refined and discerning sense of agency. When riders were passive, unclear, or disconnected, the dopaminergic systems that support purposeful movement were not recruited. But when intention, clarity, and meaning were present, his ventral striatum and basal ganglia engaged, and his movement transformed into coordinated, expressive action. This selectivity was not resistance and not shutdown; it was the signature of intact motivational circuitry and cognitive discrimination. Tuff responded to meaning, not to pressure.

Over a lifetime of consistent, low-threat, intelligible experience, his nervous system was shaped through adaptive neuroplasticity. That kind of durable learning is supported by **BDNF**–TrkB signaling, which strengthens synaptic connections and reinforces emotional resilience, motor learning, and cognitive flexibility. In other words, Tuff's composure was not merely behavioral; it was structurally embedded.

As a herd leader, he also served as a stabilizing presence for others. Calm, regulated individuals can lower arousal in those around them through **social buffering**, associated with oxytocin and other affiliative neurochemistry. Without force or dominance, Tuff reduced uncertainty in the herd simply through his grounded steadiness, helping establish a baseline of safety and regulation for the group.

Crissi's description of shared breath, rhythm, and intention aligns with what neuroscience increasingly recognizes as dyadic regulation and co-modulation. When two nervous systems engage repeatedly in a safe, predictable pattern, breathing, posture, muscle tone, and autonomic activity can become coordinated over time. This does not mean identical internal states; it means mutual influence on arousal and regulation through relational and sensory pathways. In Tuff's case, cross-species attunement created a shared field of stability in which communication became more efficient and less dependent on force or overt cues.

BRAIN-DERIVED NEUROTROPHIC FACTOR (BDNF)

A protein that helps the brain decide which neural connections to strengthen or weaken. When mature BDNF binds to TrkB, it supports learning, memory, resilience, and long-term potentiation. When pro-BDNF binds to p75NTR, it promotes long-term depression, weakening or pruning connections, especially under prolonged stress.

SOCIAL BUFFERING

Horses are deeply social animals whose nervous systems are influenced by the emotional and physiological states of those around them. *Social buffering* refers to the biological process by which the presence of a calm, familiar, or trusted companion reduces stress and supports emotional regulation. When a horse perceives safety through proximity to another regulated being, horse or human, activity in the stress system (HPA axis) decreases, cortisol levels drop, and bonding-related hormones such as oxytocin increase.

Tuff was not shut down. He was fully integrated. He embodied what becomes possible when a nervous system is shaped by consistency, clarity, predictability, and trust. His mind did not brace. His body did not panic. He did not react reflexively to the world around him, he responded with discernment, steadiness, and quiet intelligence. And because outward appearance can mislead, his case underscores the central clinical principle: careful observation must be paired with the horse's history. Without that context, a quiet horse can be misunderstood and assumed to be calm when it is not or labeled "hopeless" when recovery is still very possible.

CHAPTER 2

BLUE

The "Shut-Down" Gelding

MARK

Back in the late 1980s and throughout the 1990s, we began seeing a rash of horses at our clinics exhibiting what I'd call an overall dullness and lack of interest in engaging in anything that was being asked of them. One of these horses was a sixteen-hand quarter horse gelding named Blue.

Blue was a very handsome and well-built registered six-year-old who the owner, Terri, said she'd gotten as a long yearling. She said that, as a youngster, Blue had been a fun-loving and curious colt who spent much of his time playing with her four-year-old mare in the paddock behind her house. By the time he was two, he was already over fifteen hands tall and starting to be a little difficult to handle, not because he was doing anything "mean" but because he was so playful.

She decided to send Blue off to get started under saddle by a trainer who'd been recommended to her by a friend. He'd been with the trainer for a little over ninety days, and when he returned, the devil-may-care little horse was completely gone, replaced with the dull and disinterested one now standing next to her in the arena.

Terri mentioned she had watched the trainer work with Blue on a number of occasions and that he'd been pretty hard on the colt because, as she put it, Blue had no respect for the trainer. Apparently, the trainer spent much of the first month in the round pen chasing Blue with a whip or throwing the loop of a lariat at him, sometimes for hours, because Blue would not look at him with both eyes.

Once under saddle, Blue went back in the round pen and was chased some more in order to get all the buck completely out of him. Oftentimes during these sessions, plastic jugs full of stones were tied to the saddle to get him used to loud noises and things bouncing on his back. This took a couple weeks before Blue was apparently okay with it.

By the time the trainer got on him, Blue refused to move so the trainer would spur and whip him with the reins until he would. Once he got Blue moving, he couldn't get him stopped.

This resulted in the trainer putting Blue through an endless regimen of lateral flexion, which Terri said consisted of the trainer sitting on Blue and pulling his head to the left and right over and over and over, something he did until Terri got Blue back. According to the trainer, this was designed to "soften" Blue up and to develop the ability to do a one-rein "emergency" stop when the gelding refused to stop upon request.

Apparently, the rest of Blue's ridden work didn't go very well either. The trainer told Terri that Blue eventually had gotten so belligerent and disrespectful that he'd taken to bucking and bolting. This caused the trainer to take what he described to Terri as "drastic measures," although no specifics were given. These measures put a stop to the bucking and bolting and finally made Blue the "bombproof" horse that the trainer eventually sent back to Terri.

"He's pretty much been like this ever since," she said, gesturing toward the motionless gelding, his head and neck hanging well below his withers.

She said that not long after she got him home from the trainer, Blue seemed to become more withdrawn, eventually becoming next to impossible to get him to do almost anything under saddle. The one exception was when he'd follow another horse out on the trail.

"It's the only time he seems happy," she said, stroking him on the shoulder. "I did take him to a couple trainers and a few clinics to see if someone knew how to wake him up, but so far, nothing really worked."

I asked Terri if she could show me the work she usually does with Blue, both on the ground and under saddle. Terri obliged by using the flag she'd been holding the entire time and using it to send Blue in a circle around her, first one way, then the other. She then used the flag to disengage his hind quarters and shoulders on both sides and asked him to side pass. The more she did with him, the bigger she had to get with the flag because Blue got more sluggish with each request.

When Terri mounted up, Blue, with eyes half closed, turned his head until his nose was resting on her right boot. He stayed like that for several minutes before Terri picked up her left rein and tried to bring his head back to a neutral position. It took a substantial amount of effort on Terri's part to straighten him as he seemed to just hang on the bit, all but ignoring her request.

Once Blue was straight, she asked him to move forward. First, she squeezed with her spurs, then bumped him with her spurs, and finally kicked with her spurs. With eyes still half closed, he finally took some steps forward, only to stop after going just a few feet. Terri repeated this pattern of kicking until her and Blue were finally traveling across the arena.

Once moving, it seemed almost impossible for Terri to keep Blue in a straight line. He wandered back and forth as if he were drunk, and nothing she did to help him straighten seemed to have any effect. Turning him took quite a bit of effort on Terri's part, as did getting him to stop.

After considering everything we'd seen from Blue and taking his past training into consideration, I explained that it certainly seemed he was mentally and physically shut down. I told Terri that I didn't think more 'training" was the way to go with him and instead suggested she give him some time completely away from humans.

I recommended finding a big pasture somewhere, the bigger the better, and turning him out with other horses for as long as she could. It would be best if he could be out for no less than six months, but preferably a year or more. I told her I felt he needed some time to just be a horse so he could decompress.

I then mentioned that once she gets him home to not only get some bodywork for him, but make sure his saddle fits, have his teeth balanced with a Neuromuscular dentist, and then do nothing more than trail ride with him, especially considering how trail riding was the only thing he seemed to be okay with.

I suggested, once he's been trail ridden for a period of time—say, six months—to bring him in the arena and see if he feels any better about going forward when asked. If he does, only go forward five or six steps, stop, and get off. Then go back to trail riding. A week or two later, try the arena again. If he goes forward when asked, only go a few steps, stop, and get off. Keep repeating that pattern until he feels okay with going forward for slightly longer periods of time, building up to eventually being able to ride around the arena.

Once his forward is reestablished, then I suggested doing a complete restart with him, building his confidence and understanding literally from the ground up. I told her I felt that the worst case scenario was that he would be a fairly dependable trail horse the rest of his life. Best case scenario, he becomes the well-rounded horse she'd hoped for from the start.

* * *

A few years ago, I ran into Terri at a clinic that Jim Masterson and I were doing together here in Colorado. It had been a number of years since I'd seen her, and I almost didn't recognize her when she came up to reintroduce herself.

I asked how Blue was doing, and she said that, while she was initially disappointed in what I suggested she do with him at the clinic, she eventually did follow my advice. After about a year and a half out on pasture, Blue came back home feeling much better overall. She did nothing more than trail ride him for the next year or so, then she started him back in the arena.

It had been slow going, but she and Blue were able to reestablish forward and eventually do the restart with him. She said she was able to enlist the help of a really good, kind, and thoughtful young woman who seemed to make all the difference for Blue. While he still had his moments, albeit few and far between, she was happy to report that Blue was now a much happier and willing horse than the one I had seen at the clinic.

I told her how happy I was for her and Blue and how lucky he was to have her as his owner. Not a lot of people would be willing to spend that kind of time and effort to help a horse that was as troubled as he was.

"I had to do some real soul searching before figuring out that, ultimately, I was responsible for the condition he'd gotten into," she said. "So I was determined to make it up to him."

"Sure sounds like you did." I smiled.

CLINICAL NOTES

DR. PETERS

Blue's story would often be described by some as that of a "shut-down" horse. From the outside, that label makes a certain kind of sense. He stood motionless. His head hung low below his withers. His eyes were half closed. He showed little curiosity, little reaction to his surroundings, and almost no motivation to move under saddle. To many observers, he appeared uninterested in life.

But Blue was not simply shut-down. He was an example of something deeper, more complex, and far more revealing about the equine nervous system: learned helplessness.

Shutdown is a temporary protective state. It happens when the nervous system becomes overwhelmed and retreats into conservation mode. In a shutdown state, a horse may seem disengaged, but the brain's capacity to learn, choose, and re-engage is still intact. When safety returns, that horse can often come back online relatively quickly.

Learned helplessness is entirely different. It develops when an animal is repeatedly exposed to frightening, painful, confusing, or exhausting situations with no way to escape and no way to change the outcome. Over time, the brain stops trying. It learns, at a biological level, that effort is pointless. The nervous system doesn't just withdraw temporarily; it rewrites its expectations of the world.

Blue's early training created exactly the conditions known to produce learned helplessness. As a playful, curious colt, he had once been full of life, social connection, and movement. But after being sent to a trainer who chased, flooded, overwhelmed, restrained, punished, and confused him for months without relief, his brain was forced to adapt to what it perceived as an inescapable reality. Eventually, the only "safe" option left was to disengage from the world altogether.

Inside Blue's brain, this chronic and uncontrollable stress kept his HPA axis, the brain's primary stress system, switched on. Elevated cortisol and norepinephrine altered the functioning of his hippocampus (which is critical for learning and memory), his amygdala (which processes threat and fear), and his motivational systems in the ventral striatum and basal ganglia. The chemical messengers that normally support motivation, curiosity, and movement, especially dopamine, became suppressed. Blue did not "refuse" to try. He lost the biological ability to feel that trying had any meaning.

Even the deeper machinery of neuroplasticity was affected. Under normal conditions, the brain relies on a molecule called BDNF (brain-derived neurotrophic factor) to strengthen and support new connections between neurons. When BDNF binds to its main receptor, TrkB, it promotes growth, learning, resilience, and **long-term potentiation (LTP)**, the process that strengthens neural connections through positive experience. This is the biology that allows learning to accumulate and confidence to grow.

LONG-TERM POTENTIATION (LTP) VS. LONG-TERM DEPRESSION (LTD)

The brain changes with experience. Two processes guide that change: *long-term potentiation (LTP)* and *long-term depression (LTD)*. LTP means connections between brain cells grow stronger when experiences are safe, clear, and worth repeating. But when fear, pain, or lack of control lasts too long, LTD can dominate, and the brain may shut down curiosity, motivation, and learning. Safety and predictability push the brain toward LTP.

However, under chronic stress, BDNF is more likely to appear in its immature form, called pro-BDNF, which binds to a different receptor called p75NTR. This pathway does the opposite: it weakens connections, retracts neural branches, and promotes long-term depression (LTD), the gradual dismantling of pathways that the brain no longer believes are useful or safe. Instead of building networks for movement, exploration, and trust, Blue's brain began dismantling them. His ability to respond, to organize his movement, to stay present, and to imagine a better outcome slowly eroded. This wasn't laziness. It was a neuroanatomical change.

Another structure deep in the brain, the bed nucleus of the stria terminalis (BNST), likely played a role as well. While the amygdala handles quick, immediate fear responses, the BNST is involved in long-term, inescapable anxiety, giving the sense that threat is always present, even when nothing specific is happening. This helps explain Blue's withdrawn, disconnected state. He was not in a momentary freeze response. He was living inside a prolonged neuro-biological expectation that danger was everywhere and escape was impossible.

His wandering movement, lack of straightness, and inability to respond to clear direction were signs of this deeper disorganization. The frontal and cingulate networks of the brain, systems that normally help with attention, decision-making, and purposeful action had essentially gone offline. Blue did not "forget" how to go forward. His brain had stopped believing that forward was meaningful.

And yet, there is hope in his story.

In the original learned helplessness experiments done on dogs in the 1960s, many animals never recovered, even when escape later became possible. That is how deep this condition can go. Blue did not heal simply because time passed. He healed because his environment changed in a deeply biological way.

When Terri followed Mark's recommendation to turn him out in a large pasture with other horses, something remarkable happened at the level of his nervous system. Movement, grazing, sunlight, sleep, and natural rhythms began to stabilize his stress hormones. The presence of other horses provided social buffering, increasing oxytocin and a sense of safety. Gentle novelty and freedom restored small degrees of agency. For the first time in a long while, Blue's brain experienced a world that was predictable, non-threatening, and responsive to choice.

Slowly, the balance inside his brain began to shift again. BDNF had new opportunities to bind to TrkB rather than p75NTR. Long-term potentiation could return. Neural connections supporting curiosity and engagement could start to grow. Dopamine and serotonin gradually came back online as Blue discovered that effort could once again lead predictably to comfort and calm.

The reintroduction through low-pressure trail rides was no accident in its effectiveness. Rhythmic, forward movement, the presence of another calm horse, and the absence of arena-based triggers allowed new experiences to be written into his brain without fear. Over time, his predictions about the world changed. Effort no longer equaled harm. Connection no longer equaled confusion. Movement equaled calm emotional stability. He relearned, at a cellular level, that life could be safe.

Blue's story does not suggest that every animal will fully recover from learned helplessness. Healing is not guaranteed. It is not linear. But his journey does reveal a critical scientific truth: when chronic threat is removed and agency is carefully restored within a stable, enriched, socially supportive environment, the brain has the capacity to reorganize toward life, even after profound adversity.

CHAPTER 3:

LUCY

The Frozen Mare

MARK

I was thirteen and had been cleaning up after horses for three years, riding for two, and training for less than one. When I say training, what I was actually doing was watching things Walter did with horses and then trying to replicate them, sometimes under his watchful eye, other times on my own.

Walter was the old horseman I worked for at the time. He was a former hardscrabble rancher that was now scratching out a living buying horses nobody else wanted that he could pick up on the cheap. He would then spend time with the horses, help them smooth out any rough edges or problems they might have, and then ultimately resell them.

Looking back, most of the horses he bought were pretty straightforward in the issues they had. They might have been hard to catch, might have trouble picking up their feet for the farrier, or might not understand how to respond properly when being ridden, and so on. However, every once in a while, he would take in a horse whose issues ran a little deeper. Lucy was one of those.

Lucy was a small nondescript red mare with just the hint of a star on her forehead. Her story, as I remember, was that she had belonged to a young girl who used her mostly to just ride around the family farm with the neighbor girl. Lucy had not had any formal training, but rather, the little girl just jumped on her bareback one day and started riding her with a halter and lead rope for a bridle and reins.

Things had apparently gone fairly well for a while until the little girl put a saddle on Lucy for the first time. Apparently, even though Lucy had never had a saddle on, she appeared fine with it. The girl and the neighbor decided to go for a ride in the nearby cow pasture, and, in short order, a cow with a new calf charged them.

Lucy took off with the girl holding on to the saddle horn with both hands. Lucy ran full speed toward a fence and, upon reaching it, went left, and the little girl went right. When the girl fell off, she pulled the saddle over on Lucy's side, causing Lucy, in a full-blown panic, to try and jump a gate.

Unfortunately, by the time she got to the gate, the saddle had slipped under her belly, and the saddle horn caught the top rail of the gate as she tried to get over, pulling the saddle to her flanks and stopping her on top of the gate mid-jump. Her body ended up resting on the inverted saddle with her hind quarters hung up on one side of the gate and her front feet barely touching the ground on the other side.

It apparently took five men, a tractor with a bucket, a series of straps, and a couple hours to get Lucy off the gate. The wreck scared Lucy to the point where she became impossible to catch or handle and scared the little girl to the point where she didn't want to ride anymore anyway. So that was how Walter ultimately ended up with Lucy.

Walter did all the initial work with Lucy, spending about two months helping her feel better about being caught, led, tied, and groomed. After that, he started working on some lunging and ultimately ground driving, none of which went all that well in the beginning, but that got considerably better as time went on.

Eventually, it was time to start reintroducing the saddle, which began with nothing more than Walter hanging a saddle blanket on the fence of her pen. At first, she didn't seem all that interested in it and mostly just ignored it. But on the second day of the blanket being on the fence, she was rubbing it with her nose, grabbing it with her teeth, and tossing it around in the pen.

It was a different story when Walter first started the process of putting it on her back. She was loose in the round pen and wasn't at all interested in having him approach her with the blanket. In fact, it took a couple weeks of slow steady work before he was able to approach her with the blanket and before she'd allow him to slowly touch her with it. It was another week before he could place it on her back.

When she eventually seemed comfortable with the blanket on her back, Walter announced it was time to start working on the same process with the saddle. A few days later, Walter told me he had to go to town and that we'd start on the saddling process when he got back.

Keep in mind that, up until that point, Walter had done all the initial work himself, only having me repeat what he'd done after Lucy already seemed to be okay with what he'd been helping her with. But on this day, he said "we" would be working on the saddling process. Anytime he'd used the word "we" in the past, it usually meant I would be getting more involved in the process than I had been.

This, it would turn out, was a bit of a misunderstanding on my part. I thought he was saying I would be introducing the saddle to Lucy while he coached me through the process. So in my infinite teenage wisdom, I made the assumption that, if he thought I had enough skill to introduce the saddle while he watched, that I'd probably be okay doing it when he wasn't around.

After all, most of the time when he had me work with a horse while he watched, he never really said much anyway. Most of the time his coaching consisted of a figurative pat on the back if things went well, or he'd just let me make mistakes and then explain what I could have done better when it was over. So, I figured if that was bound to be the case anyway, why not just get started on things early?

Not long after he left for town, I took a saddle and blanket to the round pen, and sat it on the fence rail near the gate. I then brought Lucy over. As soon as she saw the saddle sitting on the fence, she snorted and backed away from it with a little more energy than I'd seen from her in a while. It took a good twenty minutes to get her past the saddle and into the pen.

Because she seemed more concerned than normal, I decided to start with something she was familiar with and see if that would help calm her down. I put a long line on her and started to do some lunging. To my surprise, as soon as I asked her to move off she bolted, charging at top speed around the pen, always shying away from the saddle.

After quite a while of her running, she finally worked her way to a walk and then a stop, head high and sides heaving. I let her stand for a few minutes then asked her to move off again by gently swinging the end of the rope. The only response I received was her raising her head a bit more, but she didn't offer to move at all.

I asked her to move a couple more times, but she wasn't interested, so I figured that was a good time to bring the saddle blanket over and see how she did with that. She stood stock-still while I first touched her with the blanket and then worked my way up to tossing it on her back. I stroked her on her neck a couple times as a reward for standing still and noticed the muscles under the skin were as tight as fiddle strings, but I didn't really give it much thought.

Seeing as how she seemed to be standing so nicely, I figured that'd be a good time to bring the saddle over. Much to my surprise, she didn't move a muscle. Other than her head being higher than normal and seeming a bit more worried, I didn't think there was any reason not to see if I could get the saddle on her. I started slowly, just touching her on the shoulder with it.

Each time I did, her whole body would respond with a small, quick jerk. At the same time, she let out a fast, shallow exhale.

After several attempts, the exhale and jerking subsided, so I started lifting the saddle as if I was going to put it on her. Lucy's already high head rose a little more each time I lifted the saddle, but eventually, I was able to gently set it on her back. She stood as if frozen to the ground, which I naturally assumed was a good thing.

Over the next several minutes, I repeated that process another three or four times without her moving once, without her head lowering, and without her letting out a breath. Admittedly, I was pretty proud of myself for having gotten the saddle on her so many times without Walter having to be there and coach me through it. Because of that, I got a little careless and took the saddle from her back a little quicker than I should.

Apparently, that was a bridge too far, and in a heartbeat, the little mare snorted, jumped sideways, and kicked at the saddle, all in one swift movement. She kicked so hard that we would later find she had broken the saddle's wooden, rawhide, and leather-covered tree. The impact drove the saddle horn into my thigh and sent me flying across the pen. I landed in a heap with the saddle on top of me and a bloodied lip from a stirrup having hit my face.

I lay there for what seemed like an eternity, bleeding on my shirt and convinced my leg was broken, until I heard Walter's truck chugging up the quarter-mile long driveway.

***

The good news was that my leg wasn't broken and my lip healed. The bad news was, in my enthusiasm to "fix" Lucy's saddling issue, I actually set her progress back by almost a year.

Lucy was my first experience with a horse that was frozen with fear, but she was not my last. The lessons I learned from her that day: her wide-eyed, statue-like stance, the rigidity of her body, her inability to breathe, all while I did things to her and not with her, are all seared into my memory. Over the years, I have apologized to her more times than I can count, mostly at night in bed when that long-ago time runs on a loop in the sleepless hours that can accompany a particularly long day.

But along with accountability, I have also learned to offer myself a little grace when it comes to horses in my life like Lucy. While taking responsibility is always important, recognizing the gratitude for the growth they allowed is equally, if not more important. Lucy helped me minimize making that same mistake with countless other horses who were in the same boat, or a similar one, as her. And for that, I am eternally grateful.

CLINICAL NOTES

DR. PETERS

Lucy was not standing calmly. She was in a **freeze** response, a survival state governed by midbrain and brainstem circuits designed to protect the animal during extreme threat. This response is mediated largely through the periaqueductal gray (PAG), amygdala, locus coeruleus, and autonomic pathways that regulate muscle tone, respiration, and cardiovascular output.

FREEZE

A classic *freeze* response is caused by extreme, immediate threat. The sympathetic nervous system surges into high alert (fight or flight), and when that level of arousal becomes unsustainable, the system flips into immobility. The body goes still, muscles may become rigid (tonic immobility) or heavy, senses may dull, and outward responses shut down.

In this state, outward stillness is actually a sign of intense internal arousal. The body becomes rigid (tonic immobility), breathing is shallow and rapid, and movement is temporarily inhibited. This is not relaxation. It is a form of protective immobility under extreme stress. Lucy's ability to initiate movement at that moment was functionally suppressed by her nervous system.

Lucy's earliest trauma involved a sensory and physical entrapment around the saddle, with pressure around the torso, the feeling of being suspended, lack of balance, pain, and loss of control. In the round pen, it is likely that the sight of the saddle, its shape, perhaps the smell of leather, the feeling of the pad approaching her back, and the lifting motion all carried strong sensory similarity to that original event. Her nervous system did not experience these as new or neutral stimuli. Instead, it interpreted them as a renewed threat, much like a traumatic flashback in humans when the conditions are similar to the original trauma.

This type of response is comparable to what is observed in post-traumatic stress reactions, where specific cues trigger the same pattern of physiological activation that occurred during the original event. The body responds not to the present moment but to the memory encoded in the nervous system.

When Lucy stood still with the saddle on her back, that stillness was not trust. It was the result of motor inhibition combined with high sympathetic activity, a state of bracing, not learning. The rigid muscle tone felt through her skin, the elevated head position, and the absence of normal exploratory behaviors were all signs that her nervous system was overwhelmed.

Learning requires engagement, curiosity, and the ability to process new information. In a freeze state, sensory input is narrowed, and the animal is operating entirely from survival circuits. No meaningful learning can occur in this condition.

The sudden kick and explosive movement that followed the rapid removal of the saddle was a classic rebound from freeze into fight/flight. The nervous system, finally exceeding its capacity to contain the stress, released the stored motor energy in a single, forceful discharge.

This situation powerfully illustrates one of the most important distinctions in equine behavior: immobility is not the same as calmness. This case illustrates why immobility cannot be assumed to indicate acceptance; it often reflects a nervous system holding high arousal in check, and a sudden change in that state can trigger an intense motor response.

True calmness is marked by soft, mobile muscles, rhythmic, deeper breathing, a lowered head and neck, engagement with the environment, and the ability to move freely when asked. Lucy displayed none of these.

CHAPTER 4:

BRIDGER

The Mustang Made Over

MARK

Back in 2003, I received a phone call from a woman inviting me to participate in a three-day colt starting competition that she was going to be producing. She explained that the competition would run from Friday to Sunday and entail several trainers starting untouched colts under saddle in front of what she expected would be a large crowd in a yet-unnamed coliseum somewhere.

The goal of the competition would be to have all the trainers get their colts ridable over the first couple days and then culminate on the third day with everybody riding their colts through an obstacle course and doing some sort of freestyle riding something or other.

I thanked her for the invitation but respectfully declined, explaining that my colt starting process usually took much longer than three days and that I personally wouldn't feel comfortable putting an untouched colt in that environment, especially for such an important part of their early training.

She countered with the list of trainers already signed on and let me know it was one of those trainers who had recommended me. She went on to explain that it would be great exposure for me, my training style and philosophy. I again thanked her but declined the invitation.

A couple years later, in 2005, I received another call from the same woman inviting me to participate in that year's competition. We basically had the same conversation we'd had the first time, ending with me again respectfully declining.

That annual competition became very popular, ultimately spawning any number of other colt starting and "makeover" competitions here in the United States. These competitions seemed to correlate with an uptick of horses at our clinics who had been part of one of those competitions and bought at the auction that followed. Some of these young horses seemed to be doing really well with their training, while others appeared to be missing substantial amounts of understanding for what was being asked of them.

One of these horses was a four-year-old mustang gelding who was the recent winner of a mustang makeover competition. From what the owner told us, this particular competition consisted of trainers taking on wild horses for one hundred days, training them, and then competing in classes that demonstrated the horse's foundation, ability to handle trail courses, and other maneuvers. The top ten competitors moved on to a freestyle-type competition.

The owners said that this gelding, whose name was Bridger, seemed much quieter than many of the other horses, and he ultimately won the competition. Because the husband and wife were looking for a good, solid horse for their daughter, Abby, and Bridger sure appeared to be that, they placed a winning bid of ten thousand dollars for him at the auction immediately following the competition.

Apparently, from what the owners said, Bridger started having trouble as soon as they got him home. He was hard to catch, fidgety when being groomed and tacked up, and difficult to mount. Once on his back, Abby could barely steer or stop him, and in the past few weeks, Abby had come off him twice when he started bucking unexpectedly.

Earlier that morning, Crissi had given Abby and Bridger a lesson in the same arena where I was working with another horse and rider. They mostly worked on developing a stop without tension. Crissi would later say that she felt Bridger had been made to stop but that he didn't understand how to give to pressure in order to willingly stop.

Stopping softly was the only thing they worked on during their session together, which lasted about forty minutes. Their session consisted primarily of Abby and Bridger walking a small circle around Crissi, a few steps at a time, then stopping. At first, Bridger would lean heavily on the bit in the stop, but as the lesson went on, they were able to help him soften and then stop.

Later that day, when it was my turn to work with Abby and Bridger, Abby's parents showed me some videos they had received from Bridger's trainer of him going through the training process. While it's always difficult to know exactly what is going on from just a few short video clips, it certainly appeared that Bridger was being flooded with information and tasks but wasn't understanding much, if any, of it.

There were videos of a sweat-soaked Bridger being spurred into worried transitions, stopping abruptly from lope, his mouth gaping open while the rider used what appeared to be a substantial amount of pressure on a leverage bit. There were videos of Bridger, nose against a fence while being spurred into a side pass one way, and then the other, and another video showing Bridger doing a transition from trot to lope.

In that clip, Bridger scrambled in the trot, going faster until he jumped into the lope. Three or four strides into the transition, he balled up as if he was going to buck, but the video cut off before we could see what, if anything, happened.

The parents also had a video they took on the last day of the actual competition. This one was of the trainer and Bridger's freestyle routine, which consisted of them doing several circles, flying lead changes, spins, and then culminating in the pair doing what appeared to be a makeshift reining pattern.

Bridger and the trainer ran at full speed from one end of the arena to the other. Then, closing in on the arena wall, the trainer would put Bridger into a hard sliding stop. They'd then wheel around and do the same thing going the other way.

After watching the videos, I turned to Abby who was standing next to Bridger, not far from where I was standing with her parents. He looked a little worried but not really any more than most horses in a relatively unfamiliar situation.

I asked Abby if the worry we were seeing from Bridger was normal, better than normal, or worse than normal. She told me it was better than normal and asked if she could get on. I told her I thought that would be okay and explained that, at least for the time being, I'd have her ride a relatively small circle around me, just as she had done with Crissi.

Bridger started fidgeting as she put her foot in the stirrup, but Abby was up and on so smooth and fast that the fidgeting never amounted to much. Equally as smoothly, she gathered up her reins and started Bridger in a forward walk around me.

We spent a few minutes recapping the work that she and Crissi had done that morning, walking a few steps then asking for a soft stop. Not surprising, Bridger initially had trouble in the stop but pretty quickly was attempting to stop a little softer. After about six or seven laps in one direction, I asked Abby to turn and go the other way. That's when the problems showed up.

Abby took up some contact on her right rein to ask for the turn, and Bridger reluctantly responded, reversing his direction. But instead of continuing on the circle, he started walking toward the other end of the arena.

The indoor arena we were in that day was quite large, measuring three hundred feet on the long end and one hundred and fifty feet on the short end. We had stayed primarily on the east end of the arena for much of the clinic that weekend, which is where we were when Bridger decided to head toward the west end.

He hadn't gone very far when he started to show more tension than he had been, so I suggested to Abby that she turn him around and bring him back. She took up a little contact on her left rein to steer him back, and that's when all the wheels came off.

Bridger stiffened his neck, dropped his head between his front legs and went to bucking. These were not little crow hops he was doing but full-blown saddle bronc-type bucks. His nose was nearly touching the ground while his hind legs were so high that it looked like he was doing a handstand.

To her credit, Abby stayed with him for about three or four jumps before getting unseated, doing a summersault in the air and landing on her feet before tucking and rolling in the dirt. She popped up relatively unscathed (I would later find out she was on the gymnastics team at school).

Bridger bucked all the way to the far end of the arena, crashing headlong into the metal rail fence before picking himself up and bucking all the way down to where we were. Once on our end, he continued bucking all the way back down to the other end, doing two more laps before smoothing out into what could only be described as a panicked gallop. It would be another four laps at a dead run before he finally slowed to a trot.

Once trotting, he continued doing laps with the occasional straight line at a diagonal across the middle of the arena, all of this while his head was high and swinging from side to side and letting out loud snorts as he went.

He eventually stopped on the far end of the arena, head still high and snorting so loudly the sound echoed off the walls and could be heard outside the arena.

I suggested that everybody leave him alone for a few minutes and let him settle down before trying to get him caught. Eventually, he did settle. His head lowered, his breathing stabilized, and he let out a full-body shake. It was then that he was okay with us catching him.

Abby and Bridger had only signed up for that one day of the clinic, and after such a big explosion on his part, I wasn't sure there would be much of an advantage to working with him anymore, especially since he seemed so much more settled after the episode.

In the subsequent conversation with Abby and her parents, it would come out that Abby had come off him six times at home, not two, as was previously thought. She hadn't said anything about the other times because she didn't want her parents to get rid of Bridger.

After seeing what Bridger was capable of, Abby's father was convinced that Bridger had to find a new home unless I thought there was any hope for him. I explained that it looked like Bridger may have been under a lot of stress during his initial training, and much of that stress was probably based on a lack of understanding for what was being done to him but being made to do it anyway.

I suggested there was a good possibility that Bridger's nervous system may have done a reset during the episode he'd just had, and now he may be receptive to someone doing a complete restart with him. The caveat there is that the restart would probably need to be done over an extended period of time, and whoever worked with him would need to be very aware of his level of understanding and stress level every step along the way.

It seemed clear from what Abby's father said that they thought they were getting a well-adjusted, finished horse for their daughter. What they were trying to avoid in the first place was an extensive project that may or may not ever become the kind of horse they were looking for to start with.

I got the distinct impression as they left the arena that afternoon that if it had been up to Abby, she would have kept Bridger and tried to work through his issues. But the reality was, he was probably going to end up in another home.

I wish I could say Bridger was just a one-off when it comes to horses we've seen where their understanding for the things being asked of them had taken a back seat to the speed at which they were done, but that just isn't the case.

I don't know what became of Bridger, but I hope it's with someone who was willing to give him the time he needed to feel better about being in the world of humans.

CLINICAL NOTES

DR. PETERS

Bridger's behavior did not emerge from defiance. It emerged from a nervous system that had been overwhelmed, confused, and pushed beyond its capacity for regulation. What appeared in the videos, the frantic movement, excessive sweat, and gaping mouth is consistent with sustained sympathetic activation. In this state, the brainstem and midbrain dominate. The horse is no longer processing information through curiosity or comprehension, but through survival circuitry. The body is preparing to flee or fight, not to learn.

When repeated over time, this pattern becomes sensitization. The brain learns that pressure is unpredictable and inescapable. The amygdala begins to associate training cues, environmental changes, and even subtle rein contact with threat. The hippocampus, normally responsible for organizing experience into context and memory, loses its ability to form coherent learning under sustained cortisol exposure. Behavior becomes fragmented, an accumulation of reactions rather than true understanding.

This helps explain why Bridger could appear relatively "quiet" and compliant in the highly structured environment of competition yet unravel once the context changed at home and in the clinic. His performance in the arena was likely supported by rapidly formed procedural patterns encoded within the basal ganglia. The basal ganglia do more than coordinate movement; they build habits and help select behaviors based on context, motivation, and expected outcomes. When training mainly strengthens motor habit loops without building emotional safety and cognitive understanding, behavior can become narrow, context-bound, and fragile.

Within that predictable setting, the learned motor sequences could be executed with apparent smoothness. But because they were not supported by a stable limbic sense of safety or broader contextual understanding, even a small deviation, like being asked to change direction, leave a familiar zone, or experience different rein pressure produced a sharp increase in uncertainty.

This surge of prediction error re-engaged threat-processing circuits in the amygdala and midbrain, overwhelming the fragile habit-based control and culminating in an explosive, survival-driven response.

The horse's brain is built to predict what happens next. When reality doesn't match expectation, the brain produces prediction error, a fast biological "mismatch" signal. In a calm horse, that signal can drive curiosity and learning. In a stressed or sensitized horse, the same mismatch is read as danger, and survival circuits take over. The basal ganglia help turn repetition into habit, so a horse may perform smoothly in one familiar setting. But if the learning is mostly habit without emotional safety and understanding, a small change can spike prediction error and unravel the behavior.

The violent bucking episode most likely represented a panic-driven motor discharge generated in the midbrain and coordinated by brainstem locomotor circuits, the same system recruited during escape from predation. There is no "thinking" present at that moment. Only survival output.

Yet what occurred afterward is equally important. Once the energy had fully released and Bridger was left alone, his nervous system began to settle. His breathing slowed. His head lowered. He performed a full-body shake, a well-recognized sign of autonomic rebalancing as the parasympathetic system re-engages and the body returns toward homeostasis. In effect, Bridger's system "reset." This reset was not psychological in a human sense; it was physiological, a reorganization of autonomic and limbic activity.

In some horses, such an event can open a doorway for true healing if what follows is slow, predictable, agency-based, and grounded in clarity rather than force. However, repeated overwhelm without adequate recovery can lead to long-term maladaptive patterns: heightened reactivity, shutdown, eventual learned helplessness, or unpredictable explosions. That risk makes the speed-driven model of many "makeover" programs particularly concerning from a neurological perspective.

Learning does not occur best under high intensity. It occurs under predictability, safety, and controlled novelty. Only when the emotional and cognitive loops are engaged alongside the motor loop does learning become fully integrated, allowing behavior to generalize beyond a single moment or environment.

Bridger's story is not about a "bad" horse. It is about a brain that was asked to perform far beyond what it understood and a body that found no other language left but panic. Given time, clarity, repetition without fear, and a handler attuned to nervous system signals rather than surface behavior, many horses like Bridger can relearn safety, trust, and regulation, not because they are "fixed," but because the brain remains remarkably adaptable when we finally give it the conditions it needs to change. This is not just training. This is applied neuroscience.

CHAPTER 5

COMET

Rearing, Reflex, and the Rewiring of Trust

DR. PETERS

The first time I saw Comet, she caught my attention immediately. A black and white paint with a long mane. Her black was as black as licorice, and her white was bleached-sheet white—a living chiaroscuro set loose on a sweeping expanse of land with a stream running through it. She moved across the open pasture with ease like the wind on long, lithe legs, far faster than the seventeen or so other horses she shared the large acreage with. She was stunning.

But I soon learned that her speed and beauty had become her burden. Her owner, frustrated by her uncatchable ways, had resorted to force. With a group of friends on ATVs, he eventually chased her into a makeshift chute and into a small pen. Comet's natural defense was flight, until that was taken from her.

When I returned from time to time, I saw the aftermath. Her owner had backed her into a corner. She reared. He grabbed a whip. When she reared again, he struck her in the face. She responded with front hooves lashing through the air—an act not of dominance, but of desperation. The next day, it escalated. More whipping. More rearing. The man stepped out of the pen, and as Comet's feet returned to the ground, he leaned over the fence and punched her squarely in the face.

I asked why. "That's for all the times she's been a pain in the ass," he said.

I didn't ask if he thought she even understood the reason for his violence. Instead, I offered to buy her on the spot. And I did.

What followed was a long process of unwinding her nervous system. Comet had become a map of trauma. She had learned, through painful repetition, that the approach of a human predicted pain or fear. Her response of rearing and pawing was no longer a choice but a neurologically embedded reflex, an overlearned pathway, now myelinated into automaticity.

She was living in a state of sympathetic overdrive with norepinephrine and cortisol flooding her system, the hypothalamus firing through the HPA axis, the amygdala lighting up with each human approach. Her internal landscape was noise, far too much arousal for the signals of communication to get through.

People warned me. Told me she was dangerous. That I was a fool. That she should be put down.

But I had no doubt about my task. I needed to lower the norepinephrine in her system and help her nervous system rewire—to reclaim choice, curiosity, and calm.

At first, just the sight of me across the round pen would cause her to shift into sympathetic arousal. So, I began from a distance. I watched for any signal that my presence triggered unease, and I stopped. That was the language we began with: stopping. Respecting thresholds. Letting her see that not every approach led to escalation.

Over time, her system began to recalibrate. I could come closer. Then inside the pen. I approached only from the side. Never directly, never with pressure. I didn't ask for control. I invited curiosity. I offered choice. I let her check me out.

There was one key moment during our work together that showed the shift. Instead of rearing, becoming light on the front end, moving away or bracing, she paused in a steady, self-regulated stance. Her muscles stayed loose, her ears flicked toward me, and she lowered her head slightly as if weighing her options. Then she chose to take a single step forward and check me out.

It wouldn't have looked like a dramatic scene for anyone watching, but it was the clear sign of a significant turning point: her arousal stayed within a tolerable range, allowing a different circuit to come online. In that moment, the old defensive loop driven by amygdala activation and conditioned defensive responses was interrupted. I could see the hint of a path where curiosity and approach behavior could begin to replace the automatic reflex of fear.

It took eight months before I could work with her like any other horse. Even then, there were traces, like a trail ride when someone approached too assertively on foot and she'd raise her front feet, but those moments faded. With each positive experience, the old reflexes weakened. New pathways formed.

Now, more than twenty years later, I still have Comet. She's traveled the country with me, working cows, climbing mountains, swimming in the ocean, bushwhacking through forest trails. The horse that people said was crazy, dangerous, and irredeemable turned out to be the best horse I've ever had.

She wasn't broken. She was hijacked by her own neurochemistry and by the repeated activation of defensive circuits. We now know that dopamine plays a role in modulating fear in the amygdala. I didn't know that at the time, but I could feel it, the moment when curiosity started to take the place of fear. When her eyes began to soften. When she chose to come toward me instead of away.

There was a great horse in there all along. She just needed the conditions to find her way back to herself.

CLINICAL NOTES

DR. PETERS

Comet's story is a powerful example of how high stress and trauma can condition a horse's nervous system into states of hyperarousal and defensive reactivity. Her repeated exposure to fear-based handling led to the development of a **conditioned autonomic response**, where the mere sight of a human became a trigger for sympathetic activation.

From a neurobiological perspective, her rearing and pawing behavior, while often labeled as dangerous or dominant, was better understood as a reflexive survival strategy encoded through fear-based learning. High levels of **norepinephrine** and cortisol, driven by the hypothalamic-pituitary-adrenal (HPA) axis, reinforced an overactive amygdala response, effectively hijacking higher-order processing in the **frontal cortex**, cingulate, and basal ganglia.

With repetition, this defensive motor pattern had become an overlearned behavior, supported by activity-dependent and increased synaptic efficiency, essentially forming a fast-access pathway within the brain's fear and motor circuitry.

In such cases, training methods that then apply additional pressure tend to

CONDITIONED AUTONOMIC RESPONSE

A *conditioned autonomic response* is a behavior practiced so often it becomes fast and almost effortless. These patterns are stored in the basal ganglia, helping a horse respond reliably without having to "think it through." With clear cues and timely release, automatic responses reduce stress and conserve mental energy. The same system can also lock in defensive habits. After rough handling or fear, a horse may automatically brace, flinch, or avoid when a hand or halter appears. With patient, low-stress repetition, new experiences can replace old ones through neuroplasticity.

NOREPINEPHRINE

Norepinephrine is a neuromodulator that shifts a horse from rest to alertness. When something demands attention, like a sudden sound or a gate opening, norepinephrine rises, sharpening focus, speeding reactions, and preparing the body to move. In moderate amounts, it supports learning by keeping the horse attentive and responsive to cues. But when norepinephrine spikes too high, the system can tip into anxiety or hypervigilance. The horse may startle easily, react defensively, and have difficulty returning to calm, making automatic "protective" responses more likely.

exacerbate the problem. Comet's healing began when approach behaviors were modulated to reduce threat, allowing her system to downregulate.

Down-regulation is the nervous system's shift from high arousal—fast heart rate, rapid breathing, and muscle tension—back toward a balanced baseline. In horses, this recovery is driven by the parasympathetic nervous system, which counteracts "fight or flight." As the system settles, you may see slower breathing, softer posture, blinking, or licking and chewing. These are common signs the body is returning toward equilibrium as stress chemistry decreases.

This approach of observing thresholds, reducing input, and offering agency began to engage the parasympathetic nervous system and allowed for neuro-plastic change (rewiring). Her curiosity returned, a sign of dopamine's regulatory role in the amygdala and the emergence of what behavioral neuroscience now calls **reward-mediated fear extinction**.

FRONTAL CORTEX

The *frontal cortex* in horses is relatively small compared to humans. Its main functions are directing attention, supporting flexible responses, and helping the horse adjust behavior in real time. Unlike humans, it is not specialized for reasoning, planning, or abstract thought. Instead, it works with memory and emotional systems to guide quick, adaptive responses and partners with the anterior cingulate cortex in decision-making.

REWARD-MEDIATED FEAR EXTINCTION

Fear extinction is when a learned fear response fades after the cue is repeated without anything bad happening—for example, a horse calmly approaching a trailer or saddle that once triggered alarm. Reward or well-timed reinforcement (including clear release of pressure) can speed this process. This is *reward-mediated fear extinction*. Dopamine helps "stamp in" the new meaning of the experience. We now have evidence that dopamine signaling within the amygdala itself can shift amygdala activity away from fear and toward learning safety and predictable reward.

Ultimately, Comet's transformation illustrates how trauma-informed, evidence-based interaction can shift the nervous system out of survival mode and into a state of relational safety and learning. With time, predictability, and choice, even deeply conditioned defensive responses can be rewired.

NEUROPLASTICITY:

Learning and Relearning

CHAPTER 6

ROCKY

The Path from High Alert to Calm and Confident

MARK

In late fall of 2006, I found myself in need of a good ranch and clinic horse. This wouldn't have been an issue in the past, considering we had been breeding, raising, and starting our own colts for years. But those days were long gone. I was now a full-time clinician who spent much of my year on the road, so raising horses and livestock at home became increasingly more difficult. So much so, that we unfortunately found that the prudent thing to do was to downsize, selling off all the horses in our breeding program.

The first person I called to see if he had any horses available was my friend, Lloyd. Lloyd was an old horseman from western Minnesota who, for decades, had been raising some pretty amazing, registered Quarter Horses. In fact, Buck, the horse who was the subject of my book *Life Lessons from a Ranch Horse*, had come from Lloyd. The stallion and most of the brood mares we had used for our breeding program, along with a handful of good ranch horses, we'd picked up from him over the years.

I called Lloyd from a clinic venue in California that late autumn day, and in his quiet, slow and thoughtful voice, he told me he did have a good gelding he'd think about selling. His name was Rocky, a seven-year-old that happened to be a nephew of Buck and who had become his go-to using horse on the ranch. I told Lloyd that I didn't want to take his good horse from him and asked if he had any others he could part with instead.

There was a long pause before he said, "You should probably take this one. I use him so much I'm not getting any of my colts rode." Due to my schedule, I couldn't get up to Minnesota until about a week before Christmas. Even then, I had a very small weather window in which to get up there and back without getting hit with one winter storm or another along the way.

Because of the limited time I had available, I wasn't able to ride Rocky once I got to Lloyd's place. Not that I would have been too enthusiastic about doing so anyway, seeing as how I had recently been in sunny, warm Southern California, and it was at least fifty degrees colder in Minnesota with a stiff wind blowing in from the north.

While sitting in Lloyd's kitchen, Lloyd's wife, Janice, drew up a bill of sale while Lloyd went through a manila envelope from which he slowly took Rocky's registration papers. He sat the document carefully in front of him on the table, gently running his hand over it as we chatted about this and that. Janice asked if I wanted something to eat or drink.

"I should probably get going," I said. "I want to try to get ahead of that storm coming down from Canada." Janice stepped behind Lloyd and placed her hand on his shoulder. He glanced up at her then signed the transfer over to me. He finished, slid the papers across the table to me, then: "If you don't mind," he forced a smile. "I'll let you load him. He'll jump right in your trailer. He's good that way. I can't watch him go."

***

Between the Christmas holidays and the high-country weather here in Colorado that year, I didn't get a chance to ride Rocky until I took him to Arizona for a clinic in January. I wasn't too surprised at Rocky's reaction to the situation, which was one of nervous alertness. His head was high, body tight, he was looking at everything seemingly all at the same time, his movements were quick and jerky, and standing still simply wasn't an option for him.

A big part of this, I'm sure, was the fact that this was completely different and unfamiliar from anything he had done on the ranch in Minnesota. But another part of it was no doubt due to the way Lloyd had ridden him for the past five years. Lloyd had told me on several occasions that he liked his horses to be "awake and alert" when he rode them. This often would translate to his horses being a little high headed and "looky."

Rocky was certainly that during the first three days of that clinic. He spent all day with his head in the air looking everywhere except the pen in which we were working. He also never had all four feet on the ground at the same time, but rather, we side passed here and there, danced in circles, spun left and right and charged around in figure eights, with me directing most, but not all, of his movement.

Student after student would come into the arena with their horse, and while working with them, I would simultaneously be asking Rocky to refocus his attention on what we were doing instead of staring at the truck out on the road, or the tractor over by the tree line, or the birds flying overhead, or the auditor going into the portable toilet, or the one drinking from her water bottle, or the one coming out of the portable toilet, or any number of other worrisome goings-on within his purview.

On the fourth day around lunch time, and after another morning of redirecting Rocky's energy and attention, I tied him to a nearby hitch rail so I could answer questions from the auditors. About fifteen minutes into the question-and-answer session, I turned around to see Rocky had laid down, still tied to the hitch rail, and was dead asleep.

That afternoon, Rocky was a completely different horse. He was able to stand still without my help, and while certain things still would get his attention, he was able to check things out and then come back on his own. As I mentioned, that was in January. We wanted Rocky to feel better than he did about his new home and job, so we decided to really take our time with him going forward. Because of that, I rode him only in the walk from January until May of that same year. We ended up getting another horse in mid-May that I was going to need to spend time with, so Crissi started riding Rocky. She rode Rocky in the walk from May until September before working into the trot. The following March, a year and three months after I had purchased him from Lloyd, we finally began loping him.

I probably should add that, when I say we rode Rocky in the walk from January till September, it's important to understand that didn't mean he was just walking around in an arena during clinics. While that was certainly a big part of what he was doing, he was also out on the trails, crossing water and bridges, opening and closing gates, ponying colts, doing ranch work, including working cattle, sorting pairs, roping and dragging calves to the fire for branding, and being hauled from one end of the country to the other countless times.

As with all our horses, when it came to Rocky's handling and training, we didn't want him to just do the things we were asking of him, we wanted him to understand the things we were asking of him. Understanding translates to willingness, which in turn translates to effortless movement. Sometimes developing that understanding can take some time, and in Rocky's case, time was something we had plenty of.

My personal goal with our horses is that, ultimately, they can function as easily with me on their back as they can on their own out in the field. By the time we started working Rocky into his trot and lope, his transitions were soft and had that effortlessness we were hoping for. Speed control within a gait or transition was easy, and most everything we did together could be done using little more than a thought as a cue.

Over time, Rocky became my go-to horse. If I needed a job done, he was the one I could always count on. During the filming of our movie *Out of the Wild*, Rocky played the lead role of a troubled mustang that eventually becomes a solid ranch mount.

Interestingly, a good deal of what he was required to do for the role was certain, very specific liberty work while inside a forty-foot pen. And while the work we had done with him under saddle up to that point was quite extensive, we had done little to no groundwork or liberty work with him, primarily because he didn't really need it. So, during filming, the director would tell me what he needed Rocky to do for an upcoming scene, and while the crew was setting up for the shot, Rocky and I would have about twenty minutes to a half hour to figure out how to do it. This included everything from Rocky appearing to run frantically around the inside of the pen and then suddenly stopping while hitting a very specific "mark" to coming out of the shadows at night and touching his nose to an actor who was kneeling in the dirt.

Every time Rocky is seen at liberty in the film, I am in the pen with him, just out of sight of the camera and giving him direction. There wasn't one time during production where filming had to be held up because Rocky didn't do what the director wanted or because he didn't hit his mark. There were even several times when crew members would ask how we were getting Rocky to do the things he was doing. This was due to the subtlety of the cues he and I had been able to work out...some of which we made up right on the spot!

One crew member even mentioned that it looked like Rocky was reading my mind. I wouldn't go that far, but Rocky was definitely able to pick up on subtle suggestions I would give him and do so relatively quickly.

There was a time when we were clinicing down in Florida and had a few days off, so we went to visit some friends. Our buddy Tim suggested we take the horses down to the ocean and ride on the beach, to which we agreed. Crissi was on her mare, Ally, I was on Rocky, and Tim rode the quarter horse gelding he had bought at a sale a few years earlier.

Rocky and Ally had never seen the expanse of an ocean with waves lapping on the shore and seemingly endless miles of sandy beach, but after a brief minute or two to acclimate, we were on our way. We rode about a mile or two when Tim said he was going to take his gelding for a run. With that, he galloped off. I checked in with Crissi to see if she minded if I went along before heading out behind Tim.

Ally was an unflappable mare and couldn't have cared less that her buddies were leaving. Rocky was a very fast horse, and we caught and passed Tim relatively quickly. I gave Rocky his head, and we raced along the shoreline at top speed, something we didn't get to do very often.

It was low tide, and the beach was littered with debris, including dead and dying jellyfish. I did my best to steer Rocky around any of the nearly transparent blobs that I could see because, even in death, they can still be quite toxic. The interesting thing was that, after directing Rocky around a number of these flat gelatinous little troublemakers, Rocky soon took it upon himself to dodge them on his own, and he did it without slowing his pace or without me having to help him.

Rocky was also pretty much unflappable, even in some of the most worrisome of situations. This could be troubled horses running at or into him during clinics, loud and unexpected noises coming out of nowhere, colts throwing themselves around while being ponied, and any number of other situations we experienced over the years.

He would effortlessly adjust to the skill level of his rider, as was evident during the filming of the movie when he helped the lead actor, a man with little to no horse experience, look good when he rode. I could put students on him during clinics so he could demonstrate the feel of internal softness between horse and rider. Or Crissi or I could get on him and think about doing something, and he would just do it.

I could go on and on regaling all of Rocky's amazing attributes, of which there were many, but I'll suffice it to say that there wasn't much we asked of him during our fifteen years together that he wasn't successful at, including learning how to jump when he was nineteen years old.

A big part of the success we had with Rocky most certainly had to do with the level of care and understanding Lloyd used when breeding all those fine horses of his. But I also don't discount the impact that taking the time during our initial year together had on building trust that ultimately would sustain a lifetime of relationship.

CLINICAL NOTES

DR. PETERS

Rocky's story shows how the horse's brain can change and adapt once it feels safe and understood. When he first arrived, his tight muscles, raised head, and restless movement reflected a nervous system on high alert. Deep within his brain, the **amygdala** and **cingulate cortex** were constantly scanning for possible danger, keeping him ready to react. Hormones such as *cortisol* and *norepinephrine* helped maintain this vigilance but made it harder for the frontal cortex and hippocampus, which handle focus and memory, to work efficiently.

When Rocky finally lay down and slept at the rail, his body and brain reached a decisive turning point. True rest signaled that his environment had finally been judged safe enough for recovery. Sleep is not passive; it is one of the brain's most active and restorative states. Far from being a period of rest alone, it is when the nervous system restores balance, consolidates memories, and fine-tunes emotional and motor control. During **slow-wave sleep**, synchronized rhythms between the hippocampus and frontal cortex stabilize new experiences by strengthening fresh neural connections. During **REM sleep**, elevated acetylcholine and

AMYGDALA

The *amygdala* acts as the brain's early warning system, rapidly evaluating sensory input for potential threat or reward. When overactive, it drives hypervigilance and reactivity. With time and safety, its influence becomes modulated by frontal and cingulate networks, allowing for calm attention rather than alarm.

CINGULATE CORTEX

The *cingulate cortex* bridges emotional and cognitive domains. The anterior cingulate supports conflict resolution, focus, and the ability to shift attention between internal and external cues. Its strengthened connectivity with the frontal cortex helps the horse pause, assess, and choose a measured response.

SLOW-WAVE SLEEP (SWS)

Also called deep sleep, *slow-wave sleep* is marked by slow, synchronized brain waves. During SWS, communication between the hippocampus (short-term memory center) and frontal cortex strengthens new connections. The brain "replays" recent experiences, stabilizing what was learned during the day. This process, known as synaptic consolidation, lays the foundation for durable learning.

reduced norepinephrine open communication among the amygdala, cingulate, and motor systems, integrating emotional tone and refining movement patterns. Together, these phases support both synaptic consolidation and emotional recalibration, a process well documented in human and animal research. Horses experience both slow-wave and REM sleep, though in shorter cycles than humans. Studies show that horses deprived of sleep show reduced focus, irritability, and even temporary collapse when REM is missed. These findings highlight that rest is as essential to mental health and learning as exercise or nutrition.

The **insula**, which links internal sensations such as heart rate and respiration with emotional awareness, likely worked with the cingulate and frontal regions to build a renewed sense of internal safety. At the same time, GABAergic inhibition dampened excessive limbic activity, reducing excitatory drive and allowing restorative transmitters such as serotonin and dopamine to reshape neural tone. Rocky's brief but deep sleep thus marked a full neurological reset from survival readiness to a state that could support learning, confidence, and connection.

The long period of walk-only training provided ideal conditions for neural plasticity. Low-arousal, rhythmic movement strengthens long-term potentiation (LTP) across sensory-motor, frontal, and cingulate networks, reinforcing stability and attentional control. Because the work was predictable and non-threatening, dopamine could reinforce successful outcomes without excessive cortisol interference, while serotonin supported mood regulation and

REM SLEEP (RAPID EYE MOVEMENT)

REM sleep, characterized by vivid dreams and high brain activity, serves a different role. It reactivates emotional and motor circuits, integrating new experiences with older memories. During REM, levels of acetylcholine rise while norepinephrine drops, allowing emotional and procedural memories to blend into existing knowledge. In both humans and animals, this balance supports emotional regulation and skill learning.

INSULA

The *insula* integrates body sensations such as heart rate, muscle tone, and respiration into emotional perception. By linking visceral feedback with context, the insula helps the horse sense safety or tension from within, promoting adaptive responses and internal regulation.

GAMMA-AMINOBUTYRIC ACID (GABA)
on page 56

Gamma-aminobutyric acid (GABA) is the primary inhibitory neurotransmitter in the brain. By dampening excessive excitation, it protects against sensory overload and anxiety. Increased GABA activity supports learning, coordination, and emotional stability by maintaining neural equilibrium.

GABA sustained inhibitory tone. This balance allowed learning to become both efficient and emotionally grounded.

Rocky's moment of deep rest reflected more than fatigue; it marked the reorganization of his brain's learning systems. During sleep, neurotransmitters such as GABA and serotonin lower stress reactivity, while dopamine helps embed positive associations. In this way, rest transformed vigilance into readiness, preparing his nervous system for trust, focus, and lasting learning.

BASAL GANGLIA

The *basal ganglia* serves as a neural hub connecting cognitive, limbic, and motor loops. Through repetition in low-stress contexts, it automates patterns of coordinated movement and emotional steadiness. When balanced, it allows actions to feel fluid and effortless movement guided more by understanding than by tension.

As Rocky adapted, communication within his brain became more integrated. The cingulate aided attention-shifting and conflict resolution; the insula allowed him to sense his own calm and respond appropriately. The **basal ganglia** then optimized the dialogue between cognitive, limbic, and motor loops, automating responses that were once effortful and freeing neural resources for awareness and problem-solving. This stage reflected a nervous system that could regulate itself—attentive yet composed, responsive without fear.

Together, these systems form the architecture of calm awareness. As arousal subsides and GABAergic inhibition stabilizes the limbic circuits, the frontal, cingulate, insular, and basal ganglia networks synchronize, transforming vigilance into voluntary engagement. This integration is the essence of mental well-being and the neurobiology of a horse at ease in both body and mind.

By the time Rocky performed liberty work for the film, his neural networks were synchronized and efficient. The frontal, striatal, and cerebellar systems translated subtle cues into precise, coordinated movement. The same horse that once could not stand still now displayed focus, confidence, and calm awareness.

Rocky's transformation illustrates the neuroscience of equine welfare, revealing how safety, time, and consistency allow the brain to reorganize toward stability and confidence. Balanced signaling among GABA, dopamine, and serotonin supported calm, curiosity, and agency, while the basal ganglia refined efficiency across motor and emotional circuits. His story reminds us that optimal mental well-being in the horse arises not only from freedom from stress but from the freedom to think, choose, and engage with the world as an active participant in their own learning.

DAISY

Mounting Block Fears

CRISSI

The day began hot and dry on the eastern Colorado ranch where I worked, necessitating a shower each evening. But a horse named Daisy made it worth every drop of sweat I spilled. She was a liver chestnut draft cross mare who stood in a small paddock with a shelter with her buddy, Gabe the mule.

When I asked Rich, the ranch foreman, why the mare and mule weren't working like every other horse at their place, he tipped his cowboy hat back on his forehead and said, "Well, they're the reason you're here."

"Tell me more," I asked, leaning against the weather-warped fence.

It turns out that Daisy had been born and raised on the ranch, as had Gabe. Gabe, however, was in his twenties and mostly used to give the ranch owners' grandkids rides when they came to visit. Daisy, however, was only six.

"No one can ride her," Rich said. "Shame, really. With her size and build, she could be a nice mare for someone."

"Why can't anyone ride her?" I asked.

The foreman went on to explain that the ranch owner and his wife handled Daisy the same way they handled all their horses; they were weaned between nine months and a year, they knew how to be caught, led, groomed, and tied. They stood well for the farrier and had regular vet checkups.

When Daisy was four, the ranch owners didn't have time to start her, so they brought in a trainer they'd heard about from a friend of theirs. They had watched this trainer work other horses and thought she'd get along with Daisy. For the first two weeks, that's what happened. The trainer seemed to be taking it slow and was complimentary about the work that had already been done.

On a warm spring day, the trainer decided it was time for her to get on. She led Daisy to a mounting block and leaned on her back, draping her arm over the other side and petting her.

"You saw all this?" I asked Rich.

"I did. Before I knew it, that gal had swung a leg over Daisy and started to kick her. To her credit, at first, Daisy didn't do anything, but then this trainer turned Daisy's head with the lead rope and gave her a good wallop with the end of it, and all hell broke loose."

Daisy had, at first, tried to run, but confined by the round pen walls, she started bucking. The trainer came off, picked up her hat, and got Daisy stopped and back over to the mounting block.

"I tried to tell her that she didn't need to get back on. Didn't seem like it was going to get any better."

"Did she stop?"

Rich shook his head. "No. The same thing happened three more times before the trainer limped out of the pen. I let the ranch owners know about it, and they invited her to not come back."

I asked what happened after that. Rich said that the owners started all over with the mare, taking their time. In the fall, when they turned the herd out on winter pasture, Daisy seemed like she was back to her old self.

The next spring, when he tried to work with her, Daisy couldn't be led into the round pen at first, and when they eventually got her there, she wouldn't stop running. "Couldn't even get near the mounting block, much less get on it," the foreman said. That's when Gabe came to live with Daisy, and Daisy was worked with when he or the ranch owners had a few moments. They could lead her into the round pen and around inside, but again, she wouldn't get near the mounting block. They even bought another one in a different color, hoping that she wouldn't associate a new one with an old experience.

"Well, I'm looking forward to working with Daisy tomorrow," I said, shaking Rich's hand.

The next morning, after catching Daisy, I tied her to the hitch rail and began grooming her. I didn't want to jump right into the thing that worried her the most until we had a sense of each other. I led her around the outside of the pen we'd be working in, then led her in and out, watching for any signs of worry. Besides her head going higher and her breath louder, she followed me willingly.

Daisy and I repeated the routine for the next two days, each time she carried her head lower for longer, and being in the round pen wasn't worrying her like it did on the first day. I turned and closed the gate and unclipped the lead rope from her halter. She looked over her muscular shoulder at me then ambled off, searching the edges for stray blades of grass.

As she did that, I went out of the pen and grabbed the red mounting block, propping it on my hip as I carried it in one hand, and opened and closed the gate with the other. When I set it down with a hollow thunk, Daisy raised her head, snorted, and walked away. I sat down on the block, watching as she paced the fence, her head turned away. Debating whether I needed to leave and let her work it out on her own, I decided to stay and see if she could find a way to settle.

Twenty minutes of mostly trotting at the edge of the round pen went by before she stopped and looked at me. I waited. She put her head down and blew at the sand, then she shook her neck. I got up and moved the mounting block away from the gate then left the pen.

Once I was out, I was surprised to see her make her way over to the plastic block and, after a few snorts, push it over on its side. She came over to the gate, her ears forward. I figured that was a good start and led her back to her pen as Gabe brayed a welcome.

The next morning, I flashed on the memory of Daisy pushing the mounting block over. She was behind me as I led her into the round pen and closed the gate. I tipped the mounting block right side up, and without thinking, stepped up onto it, Daisy snorting behind me. The rope was taut in my hand as she leaned back, her eyes wide. I noticed she wasn't breathing.

Knowing I'd made a careless choice by getting onto the mounting block too soon, I also didn't want to get off it while she was teetering between staying and fleeing. Daisy made the choice for me; with a flick of her head, she sat back on her hindquarters and the lead rope slipped out of my hand. She trotted to the farthest point away from me and shook her head, blowing hard through her nose. I exhaled almost as loudly, wondering how far my mistake had set us back.

After going to Daisy and gathering up the rope, I led her toward the mounting block once more, then we turned right and made a large circle around it. I listened as her hoof beats behind me went from high-pitched (usually a sign of agitation, and at the very least, tension) to a slower and low pitch. As we reversed and went the other way, she began to breathe, and the lead rope was slack in my hand. I stopped for a moment, turning toward her. She flicked an ear in my direction as she looked at the mounting block, raising and lowering her head.

I unclipped the lead rope and left the pen, taking a drink of water and watching Daisy, repeating what she did the day before, inch toward the mounting block, nose lowered, and knocked it over, a shudder going through her body when it thudded to the ground.

This time when I entered the round pen, I stood the mounting block upright before clipping the lead rope on to Daisy's halter. I turned and looked at her, running through several options in my head: I could put the lead line on and circle the mounting block again, seeing how close we could get, or I could attach a longe line and get on the mounting block and see how far away Daisy needed to be, then help her feel better about coming closer.

I decided on using a longe line, which would give Daisy the choice to stay further away as I stood on the mounting block. After attaching the worn cotton line, I fed the rope out as I walked away from her, then I stepped up on to the block. Daisy snorted and trotted around me before settling at a walk, one eye and ear tilted in my direction, the rest of her body leaning away from me. I watched as she drifted to a halt in front of me, raising and lowering her head.

At this point, it would've been tempting to put a little pressure on the line and see if she could step closer, but remembering my mistake earlier, I waited instead. I watched the clouds drifting across the blue sky, and heard a crow in the distance. I felt the warm sun on my back. When I looked at Daisy again, her neck and head were lower and the muscles in her body relaxed. With a sigh she cocked her hind leg, closed her eyes, and licked and chewed. I stepped off the block and led her out of the pen.

The next three months progressed smoothly between Daisy and me. After a couple of sessions with the longe line and me on the block, she began to close the distance between where I stood and her own body. We began showing her how to orient her body perpendicular to my body, mimicking the position she would need to be in to have a rider get on to her back. At first, she could be in this position eight feet away from me. During our time together, it was Daisy who closed that gap between us, until in the second month she could line up to the mounting block, and I could pet her neck. Her ears were forward, she was licking and chewing, and she would turn her nose toward me, gently touching my leg.

The next month we spent petting her body, draping my arm over her back, leaning my bodyweight against her, and then finally going through the whole process again with a saddle on. There was a brief setback, but once we repeated what we'd been working on, she seemed to grow calmer.

By the end of the third month, I could lead her to the mounting block where she would line herself up, I could put my foot in the stirrup and mount up. Daisy's head stayed down, her body relaxed, and I could feel her breathing underneath me.

Rich was watching the day this happened. I beamed at him from Daisy's back before I dismounted and led her to where he was standing.

"Funny how it took three months to correct a problem that lasted less than five minutes," he said.

I nodded. "Only if you weren't Daisy," I said.

The rest of that summer I spent with Daisy was easy. We progressed from riding her in the round pen, making sure she understood how to stop, turn, and back. We rode in and out of the round pen and along the outside of it. We ponied her on the trails while she was saddled, which then led into me riding her as she was ponied, which turned into me riding her out on the trails.

When I left the ranch, driving away and seeing Daisy in my rearview mirror was one of the more difficult moments I'd had. Mixed in with that sadness was the satisfaction of knowing that we were able to help her overcome her worries and hopefully lead a productive life as a ranch horse. Since Daisy, I've had many moments to reflect on the lessons she taught me. The lesson I remember the most has been that I've never once regretted choosing to go slowly.

CLINICAL NOTES

DR. PETERS

Daisy's mounting block troubles illustrate how associative memory within the equine **limbic system** can transform a neutral object into a signal of danger. The initial training incident paired pain and helplessness with the sight and sound of the mounting block, forming a conditioned fear response encoded primarily in the amygdala–hippocampal network. Each re-exposure triggered sympathetic arousal, with elevated heart rate, shallow breathing, and agitation, the physiological echo of that experience.

Recovery required the formation of new neural associations rather than the erasure of the old one. Through extinction learning, Daisy's brain gradually developed inhibitory pathways that reduced amygdala activation when safety cues were present. The original fear trace remained, but its expression weakened as new experiences of calm and control accumulated, explaining why fear can resurface under stress, a phenomenon known as spontaneous recovery.

LIMBIC SYSTEM

The *limbic system* governs emotion, motivation, and memory integration. Its key structures include the amygdala, hippocampus, and cingulate cortex that evaluate sensory input for emotional significance and determine whether an experience feels safe or threatening.In horses, balanced limbic activity supports calm learning and social bonding. When stress is high, the amygdala dominates; when safety is restored, the hippocampus and cingulate help consolidate memory and guide adaptive behavior.

POSITIVE PREDICTION ERROR

A *positive prediction error* occurs when the brain experiences something better than it anticipated. In that moment, dopamine neurons fire to signal surprise and update the brain's expectations. This process strengthens learning by marking the event as important and worth remembering.

Recent research reveals that dopaminergic signaling within the amygdala itself helps mediate this process. When a once-threatening cue is encountered repeatedly without harm, dopamine neurons projecting from the ventral tegmental area (VTA) to specific amygdala circuits fire in response to the absence of danger. This activity signals a **positive prediction error**, the nervous system's way of encoding that the outcome was better than expected. In the context of **fear extinction**, the absence of harm functions as a kind of reward, reinforcing new learning that the cue no longer predicts threat. This is the same computational logic that underlies curiosity: when safety is assured, the dopaminergic system rewards exploration and approach behavior. In Daisy's case, each uneventful, voluntary interaction with the mounting block likely engaged this circuitry, allowing safety and curiosity to gradually replace fear.

Crucially, Daisy's progress depended on predictability and agency. By allowing her to approach and investigate the mounting block on her own terms, Crissi shifted activation from threat-related circuits toward dopamine and serotonin systems that support curiosity, motivation, and emotional regulation. Repeated calm exposures reinforced connections within the basal ganglia and ventral striatum, consolidating learning through reward and habit circuitry.

Physiologically, Daisy's lowered head, rhythmic breathing, licking, and soft eyes reflected activation of the **parasympathetic nervous system**, mediated through increased vagal tone and likely accompanied by **neuropeptide** activity associated with safety and affiliation, such as oxytocin. These patterns represent the biological hallmarks of trust and relaxation.

FEAR EXTINCTION

Fear extinction is the process by which the brain learns that a once-scary situation is now safe. It doesn't erase the original fear memory, instead it layers a new memory on top. For example, if a horse once panicked during loading but is repeatedly loaded without incident, the brain begins to update its expectations.

PARASYMPATHETIC NERVOUS SYSTEM (PNS)

The *parasympathetic nervous system* is one branch of the autonomic nervous system, responsible for maintaining the body's resting state and supporting recovery after arousal or stress. It slows heart rate, lowers blood pressure, stimulates digestion, and promotes relaxation and repair. The PNS works in opposition to the sympathetic nervous system, which prepares the body for action. Together, these systems maintain internal balance, or homeostasis, regulating the horse's physiological and emotional responses to its environment.

NEUROPEPTIDES

Neuropeptides are small protein-like molecules that act as long-lasting messengers in the nervous system. Unlike fast neurotransmitters that act only across synapses, neuropeptides can influence larger networks of neurons over longer time spans, shaping emotion, motivation, and stress regulation.

Ultimately, Daisy's recovery demonstrates the nervous system's remarkable capacity for neuroplastic change when given time, safety, and consistency. Her story reminds us that healing from fear is not achieved through suppression or desensitization alone, but through re-establishing the horse's felt sense of control, curiosity, and security within a predictable, relational environment.

CHAPTER 8

SUGAR

Trailer Loading Confusion

CRISSI

I shoved my gloved hands deeper into the pockets of my jacket and tucked my chin into the scarf wrapped around my neck. Watching as Logan led his dun mare, Sugar, toward the trailer, I noticed that I couldn't see the fog of Sugar's breath leaving her nostrils. The closer Logan got to where I stood by the open stock trailer, the slower the mare walked and the higher her head went up. One minute she was walking behind Logan, the next she was rearing and walking backward on her hind legs.

"Let's stop right there, Logan," I said.

He threw me a doubtful glance but did as I asked. Sugar stood at the end of her lead rope, not quite pulling away, but the slack was gone, and Logan was holding on to it with both hands. It wouldn't take much to convince Sugar that leaving was more beneficial than staying.

"Is that what usually happens?" I asked.

He nodded and adjusted his stained felt cowboy hat on his head.

"I can get her in eventually, but it's always a fight. My dad warned me about mares."

"Well, I don't know that it has much to do with being a mare, more than it looks like she doesn't understand the job when it comes to trailer loading."

"I haul her to the roping arena at least once a month. How can she not understand?"

"That's a valid question. I'll answer with one of my own. Is loading her in the trailer always like this?"

"Pretty much. I bought her a year ago, and it hasn't gotten any better."

"How long does it take to get her in the trailer?"

Logan kicked at the frozen ground. "About an hour. I run her around a lot outside the trailer, so she learns it's more work than getting in."

"Does she load after that?"

"I have to pull her in, but yeah. There's usually someone at the trailer door, and we shut it, then I climb out the emergency door."

"It sounds like what she understands is that she approaches the trailer, she runs for an hour in a circle, and then you pull her in and close the door. Does that sound right?"

"When you put it that way, I guess so."

"What would you like her to understand instead?"

"Not to be so difficult!"

"That is what you don't want her to do. What are you looking for?" He paused, took a breath, rubbing his chin.

"That I walk her up to the trailer, and she follows me in and stands quiet, and I can walk out the back and close the door without her trying to run out again."

"Great. That's our goal today."

"Today? I thought you were here for a couple of hours?"

"I'm here for as long as it takes. We're on Sugar's schedule now. Would you mind if I tried?"

"Go for it," Logan smiled as he handed me the cotton lead rope.

I stepped toward the mare to put some slack in the line and waited until I could see the twin streams of foggy breath coming out of her nostrils. We turned away from the trailer and walked down the driveway, me listening to her breathing, waiting until it was regular and not ragged.

When we turned toward the trailer, I stopped as soon as I noticed the silence behind me. I waited, listening to the tap dance of hooves as she paced one way and then another. When I turned, I saw her head was high, and the whites of her eyes were showing. Logan walked over to where I was standing.

"What are you waiting for? She doesn't rear up until you get closer."

"I'm waiting for her to feel better. Right now, she's scared. I don't want to take that state of mind closer to what she's scared of."

"What do you mean?"

I could tell Logan was asking the question because he didn't understand. It struck me that both the horse and her owner were in similar states of confusion.

"Well, you said you wanted her to follow you into the trailer and be calm, right?"

"Yes."

"That calm state of mind starts right now, this far away. If I walk her toward the trailer when she's anxious about it, all I'm doing is reinforcing the fear. We want to reinforce calm."

"I never thought of it that way," Logan said. "Would you mind if I did this instead?"

That's not a question most horse owners dealing with trailer loading issues ask. Added to this, I usually started the process myself so that we could start with a degree of clarity; a confused horse with a confused and/or afraid and/or frustrated owner isn't usually a good combination for learning a new skill. Logan, I could tell, was curious about what we were doing. Enough that he wanted to be a part of it.

"Tell you what," I said. "Let me do a little bit, and then I'll give her to you, and you can take it from there."

"That works. Can I talk to you while you work?"

"Sure!" I looked at the mare, who was now standing with her head down, her eyes alert and her ears forward.

Turning, I began walking toward the trailer again, stopping after ten feet when I could no longer hear her breath.

"Why'd you stop?" Logan asked.

"Because she stopped breathing. It's one of the signs she's nervous."

Logan was quiet, watching Sugar as she paced back and forth again, this time stopping after less than a minute.

"We're going to take her away." I could see a question forming in his eyes and took a stab at answering it. "We're doing this because she's calm, and we want to reward that state of mind."

Logan nodded and followed Sugar and me as we walked away from the trailer.

Over the next forty-five minutes, we repeated this pattern, gradually getting closer to the back of the stock trailer with the mare. My goal was to not put her in a position to feel like she had to rear again.

"Your turn." I handed the lead rope back to him. "Let's walk her away then bring her back. Remember to pay attention to her breathing"

I watched for the next fifteen minutes as Sugar not only stayed calm, and Logan stopped when she wasn't, but they ended up getting within eight feet of the back of the trailer.

"This is a great place to give her a break. Let's tie her to the hitch rail, and I'm going to go warm up in my car."

I followed Logan to the hitch rail where he tied her.

"You don't have to sit in your car," he said. "Our tack room is heated. And there's coffee."

Not normally a coffee drinker, I accepted the chipped mug, wrapping both of my hands around it to absorb the warmth.

"Do you have any questions while we're here?" I asked.

"Why didn't we keep going? She seemed calm enough."

"We need to give her some time to absorb what she learned. If we would've kept pushing her, that calm state of mind would've gone right out the window."

We let ten more minutes go by as we finished our coffee, then we went back out. Logan untied Sugar and walked toward the trailer with me following them.

This time, they got within four feet of the back of the trailer before stopping.

"That was good timing."

"What do we do now?"

"Wait a little bit. Let's see if she can relax."

"Do I need to make her run in a circle?"

"We don't need to do that."

Logan nodded, watching as the mare lowered her head to the ground, running her muzzle over frozen weeds.

"Let's walk her away and bring her back and see how she feels."

This time, both Sugar and Logan got to the back of the trailer. She lowered her head again and blew against the mats, running her muzzle back and forth before backing up.

"Let her go. She's creating space so she can figure out what she just sensed."

The mare stopped, lowered her head, and shook her entire body, the tremor running from her muzzle to the end of her thrashing tail. She let out a sigh, blew through her nose, then closed her eyes.

"What now?" Logan asked.

"We wait."

"Is that your answer for everything?" It was the first sign I'd seen of humor from this serious young man.

"Most things," I grinned back. "I never regretted going slow, but I have many regrets about going too fast."

"I guess you're going to tell me to walk her away again."

"Actually, I wouldn't mind seeing her go toward the trailer."

When they reached the back of the trailer, she put her nose down again and ran her muzzle along the black mats. This time, she didn't rush backward.

"Should I try and get her in?"

"Nope. Not yet."

"Let me guess. You want us to wait."

This time, it was my turn to laugh. "You're right."

By giving her time near the trailer in a quieter frame of mind, we're reinforcing that the trailer isn't as scary as she thought it was. The more time

Sugar is calm, the more the chance we have of asking her to follow Logan inside and have her be okay.

We repeated this pattern several more times, the dun mare eventually standing with her neck and head inside the trailer, eyes half closed. We walked her away and chatted for several minutes, then Logan led her back.

"This time, I'd like you to step up inside the trailer and see what she does. Is there a cue you use to ask her to go forward?"

"Just a cluck."

"Do that as you step up."

He stepped into the stock trailer, and I heard him clucking. The mare stopped, rocking back and forth.

"Stay right there. Keep your hand relaxed, and remember, we aren't pulling her in."

He nodded, then clucked again, putting a little pressure on the lead rope. Sugar raised one front leg and pawed the bumper of the trailer, shooting backward at the sound her hoof made against the hard rubber. Logan went with her, waiting until she stopped, then he led her a few steps forward.

"That was a great choice, Logan. Let's give a minute and try again."

I lost track of time as we worked with Sugar, watching as her pawing went from hitting the bumper, to raising her leg high enough to place her hoof on the mat, to pushing off her hindquarters and standing inside the trailer with one trembling front leg. The sun was watery in the gray sky, and I guessed it was probably close to noon.

"Time for another break."

"I could use one too," he smiled.

Once we were in the tack room, I said, "I'm tempted to come back tomorrow and leave what we did today on a good note."

"She's so close to getting in, though!"

"She is. This is where it's tempting to speed things up and get what we want."

"Yes." He sat for a moment, hands held out in front of the space heater. "I don't have to haul her until the day after tomorrow, so I guess if you came back, that would work."

"Great. Go ahead and put Sugar away, and I'll see you tomorrow morning."

The next morning, we met by the trailer. The sky was clear, and although I was still bundled up, I could tell it was going to be warmer. I watched as Logan stroked Sugar's neck, and she stood, once again, with her head inside of the trailer, eyes half closed.

"All right, let's see what we've got."

He walked her away, then he turned and upon reaching the trailer, stepped up, clucking.

She raised one front leg, and then the other, standing with her front end in, running her muzzle along the metal slats, then touching the ceiling.

"Go ahead and gently ask her to back out."

"How do I do that?"

"Just a light pressure on the lead rope. You can say 'back' if you want."

We repeated this pattern for the next half hour before taking a break. This time, it was Logan asking me if he could take her away for a bit.

"You read my mind," I smiled.

After another break, we practiced helping Sugar load her front end in and back out. As I watched Logan working with his mare, I was glad I'd let him do the bulk of the work. Despite their misunderstandings about trailer loading, I could see they trusted each other.

Logan led the mare over to me. "I have to be honest about something."

"What's that?"

"The reason I'm hauling her tomorrow is so that a friend of mine at another ranch can try her out. Her family uses their horses mostly on the ranch doing cattle work."

"Oh. That sounds like a good plan."

"But..." he began. "The reason I was letting my friend try her was because I was so tired of having this trailer loading be such a big deal. Sugar is great at everything, except this."

"From where I stand, it looks like she could be great at this too."

He nodded. "I'm thinking the same thing. I may have to back out of giving her to my friend. There aren't many horses I've had that I like so much."

We chatted for a few more minutes, and after Logan had led Sugar into the trailer with her front end, I asked him to cluck and put light pressure on the lead rope and see what she did.

"Aren't you worried she'll rear up again?" he asked.

"I don't think she will."

The mare stood quietly, her front end in the trailer, her hind legs only inches away from the back of the trailer.

He nodded, clucked, and put light pressure on the rope, stepping backward.

Sugar raised her head then lowered it and jumped in with both hind feet at once.

"She's always an over-achiever," Logan laughed. "What now?"

"Let her stand for a moment. I don't think she's too worried."

Besides running her muzzle along the floor and the metal sides of the trailer, she stood still.

"Okay, let's ask her to back up." I wanted to be sure he offered his mare a little support on the lead rope as she backed out, to make sure she didn't rush and scare herself.

There was a bit of scrambling as the mare backed out, but as she walked behind Logan away from the trailer, she blew through her nose, shook her neck, stopped, and shook her whole body.

"That was amazing!" Logan said. "I'm definitely keeping her."

At the end of the next hour, Logan and the dun mare could walk into the stock trailer together, she could stand tied quietly, and he could shut the door.

"I don't think you need me anymore," I said. "When you come out tomorrow, repeat this whole process, and gradually lengthen the time the door is shut before you lead her out."

"Okay," he said. "Does it matter if she backs out or follows me out?"

"Good question. If it were my horse, I'd show her both. That way it won't worry her if she gets in a trailer where the only option is to back out."

"Then we'll do that," he said, shaking my hand. "Thanks for helping us."

The sun was unimpeded by clouds, the winter blue sky framing Logan and Sugar as they walked back to the paddock, his hand resting on her neck.

CLINICAL NOTES

DR. PETERS

Sugar's trailer-loading pattern can be understood through the function of the basal ganglia, which organize and reinforce learned behaviors through interconnected loops. The cognitive loop, involving the **caudate nucleus**, is responsible for forming predictions and expectations about what typically happens in familiar situations. For Sugar, the caudate had encoded the approach to the trailer as the beginning of a confusing and stressful sequence. Her brain anticipated difficulty before it occurred, not because of personality or willfulness, but because past experience had shaped her expectation.

The motor loop, centered in the **putamen**, executes and stabilizes movement patterns that have been repeated enough to become automatic. Sugar's rearing, backing, and bracing were not improvisations; they were rehearsed motor responses stored as procedural memory. When the situation resembled past loading attempts, the putamen simply activated the same movement pattern it had previously been asked to use.

CAUDATE NUCLEUS

The *caudate nucleus* is part of the basal ganglia's cognitive loop, linking sensory input with decision-making and goal-directed behavior. It helps the horse evaluate actions, predict outcomes, and adjust responses based on prior experience.

PUTAMEN

The *putamen* forms the motor loop of the basal ganglia, translating intention into coordinated movement. It refines motor patterns through repetition, allowing actions to become smooth, efficient, and eventually automatic.

VENTRAL STRIATUM

The *ventral striatum*, which includes the nucleus accumbens, is the brain's hub for linking motivation with action. It integrates emotional signals from the limbic system with dopaminergic input from the midbrain to evaluate rewards and drive goal-directed behavior.

The emotional or limbic loop, which involves the **ventral striatum** (including the nucleus accumbens), assigns emotional value to experiences. Because earlier loading episodes had involved pressure, escalation, and uncertainty, the ventral striatum had classified the trailer as something to avoid. Once an emotional valuation like this is in place, it strongly influences both the cognitive expectation and the motor response. In other words, Sugar did not just think the trailer would be difficult; her nervous system felt that it was unsafe.

The change seen in these sessions occurred because the emotional loop was addressed first. By slowing down, stepping away when tension increased, and reinforcing calm states at a distance, the ventral striatum began to reassign the trailer from a negative to a neutral (and eventually workable) stimulus. Once the emotional valuation shifted, the cognitive loop could update its prediction: approaching the trailer no longer automatically signaled conflict. Only then could the motor loop begin to develop a new behavioral sequence of approach, remain regulated, explore, and step forward.

This process is an example of neuroplasticity in **procedural learning**. The behavior changed not through force, fatigue, or correction but through creating conditions in which Sugar's nervous system could reorganize the pattern at its roots. When calm, the basal ganglia can revise both the emotional relevance of an experience and the motor actions associated with it. When stressed, previously established avoidance patterns simply re-emerge. What was reshaped here was not just movement, but the underlying expectation that drives movement. Sugar did not merely learn how to load; she learned that the trailer no longer predicts distress. That change is durable because it is stored in the same neural systems that formed the original habit.

PROCEDURAL LEARNING

Procedural learning is the process by which the brain encodes motor skills through repetition and practice. In horses, it allows responses such as transitions, lateral movements, or trailer loading to become automatic over time. This form of learning depends on the basal ganglia, cerebellum, and motor cortex, which refine movement patterns through consistent feedback. With clear, repeated cues and appropriate timing, procedural memory replaces effortful control with fluid, automatic motor patterns.

CHAPTER 9:

DOMINO

Trained with Treats

CRISSI

I watched as a tall woman led a brown and white Appaloosa horse into the arena. She stopped, reached in her pocket, and fed a treat to the horse before walking up to me, smiling.

"Hi!" she said. "I'm Olivia, and this is Domino."

I shook her hand and introduced myself, then I asked what she'd like to work on.

"Well," she said, stroking the gelding's neck. "I bought him from a woman who was trying to do dressage with him, but he was either bucking or didn't go forward, so she sold him to me."

"What's his job now?"

Olivia reached in her pocket and fed Domino another treat before answering me.

"I'd like to do some Reining, and maybe some Western Dressage. Right now, we go out on the trail a lot. He seems to like it."

"Has he had any more bucking or not going forward episodes with you?"

Olivia shook her head. "No, the trainer I'm working with recommended that I have some bodywork done for him, and the vet examined him too. She gave him an injection that he gets once a month for his joints. The bodyworker also visits once a month."

"That sounds great. What does he eat?" Besides nudging at Olivia's pockets, Domino's body was tense, every muscle tensed toward the hand that dipped in and out of her pocket.

"I only feed him grass hay and less than a cup of low starch grain so he can get his supplements. He lives in a big paddock with my mare, where they have access to a run-in shed."

I smiled back at her. "Sounds like you're doing everything you can to set him up for success."

She fed him another treat then asked him to back up. "I've had him for less than a year, but I really like him."

"How can I help you today?"

"Well, when I began riding him, my trainer wanted to change the arena experience to a good one, so she recommended we try positive reinforcement with treats. But when he started feeling happier about working with me, and we tried to stop the treats, he wasn't as cooperative."

"What did being less cooperative look like?"

"He didn't buck, but he didn't want to go forward. Once we began giving him treats, he seemed a lot happier. But if I want to ride a reining pattern, I can't stop and give him a treat."

"That makes sense. Where would you like to start?"

"Why don't I show you what we do, and then you can let me know how to change it."

I watched as Olivia led Domino over to the mounting block and gave him a treat. She gave him another treat after she mounted, then she asked him to walk forward. He walked a half circle and stopped, bending his neck to the left so he could get a treat. She asked him to trot as he was still chewing, going for another half circle before he put on the brakes, and she gave him another treat. She changed directions and gave him a treat while he was walking, then she asked for a lope, which he went into without any trouble. He stopped after a half circle, bending his neck to the right, and got a treat. She rode over to me. I watched as the gelding stretched his neck out, his muzzle pointed at my hands, which I'd placed in my vest pockets. I stepped to the side and asked Olivia to stop her gelding away from me.

"See what I mean?" she asked.

I nodded. "You're right that he seems happy enough to do what you're asking. Do you go through a similar routine out on the trail?"

"Not really. He gets a treat when I mount up, but otherwise, he's happy to go. I'm worried I've ruined him. Maybe we shouldn't have used treats." Olivia petted his neck while he stood quietly.

"Well, from what I've seen, I wouldn't say that. We just have a little misunderstanding."

"So how do we clear it up?"

"Before we work on that, can you tell me if you and your trainer stopped the treats completely or gave them less frequently?"

"We stopped completely because we both wanted to see what he'd do."

"Did he refuse to go forward?"

"Exactly."

As we stood there, the waning sunshine warming my back, the wind lifting Domino's black mane, I spoke with Olivia about decreasing the frequency of the treats while riding and getting ahead of the pattern they'd created by doing a half circle then giving a treat.

"For starters, we're going to not ride in a circle anymore." I walked over to the side of the arena where I pulled out a ground pole and placed it perpendicular to the pipe rail fence enclosing the arena. I reached through the fence and pulled another pole to the opposite end of the area we were working in.

"We're going to ride in straight lines, going over these poles each time. If I see Domino thinking about stopping, I'll tell you, and it's your job to give him a forward cue."

"So I can use my legs?"

"Yes, or a cluck, or tapping your chaps with the end of the reins. Use whatever you need to keep the forward momentum."

Olivia nodded and rode toward the first pole. I watched as Domino stepped over it, noticing a subtle slowing of the rhythm of his walk.

"Let's ask him to move forward now."

She gave him a squeeze with her leg and clucked, and he went forward then slowed again.

"One more time, Olivia."

She applied pressure with her leg, clucked, then used her split reins to tap against her chaps. Domino pinned his ears but shot forward and maintained his walk to the next pole. His walk slowed ever so slightly, and before I could ask Olivia to give him a cue, she used her legs and clucked. After he had stepped over the pole and was moving forward, I asked her to stop Domino.

I walked toward them, staying out of reach of the Appy's curious nose. When he didn't get a treat from me, he swung his head to the left, looking for a treat from Olivia.

"Go ahead and use the reins to ask him to straighten out."

She did, and I watched as the gelding swung his head to the right, and Olivia straightened him out.

"Do I give him a treat?" she asked.

"Not yet. What we're waiting for is for a calmer state of mind. Not only do we want to give treats with less frequency, but we also want to show him that a quieter state of mind is what we reward. Eventually, we'd like the quiet state of mind, him being relaxed, to be the treat."

"I wouldn't give him treats at all?"

"Well, that's up to you. I see treats a lot like most other equipment we use with horses. It's not the equipment that's good or bad…it's how it's used."

I watched as Domino continued to try to swing his head left or right, his body rigid, an occasional tail swish or stamp of his hoof punctuating his efforts. We waited another few minutes until he lowered his head, sighing.

Olivia leaned forward in the saddle and offered him a treat.

"When you ask him to go forward, we aren't going to go in a straight line."

"Okay," she said.

"Let's tip his nose to one side or the other and then ask. We're going to use his imbalance to help him forward and hopefully take his mind off a treat."

Olivia did as I asked, and the gelding moved forward, one hesitant hoof at a time. Olivia used her legs and a cluck and rode toward a pole.

The next thirty minutes played out in the same way, Domino swinging his head less each time they stopped, and his body relaxing sooner. His ears were forward as he walked, and I changed the position of the poles, putting them closer together, adding more, then toward the end of our lesson, removing them and asking Olivia to ride in a large circle, being sure she was the one who asked him to stop, instead of him stopping himself.

By the end of our time together, they could walk, jog, and lope a large circle, reverse and do the same thing. Olivia was smiling as she dismounted and led her gelding toward me. She asked him to stop before they reached us, and Domino immediately cocked a hind leg and closed his eyes.

"That was so simple," she exclaimed.

"Do you have any questions?"

She shook her head. "I'm clear about how to keep going. I think my trainer's going to be relieved as well. She won't be spending as much money for treats, and neither will I!"

We both laughed, and I stepped toward Domino and touched him on his shoulder. He flicked an ear but didn't move.

"He's a really nice horse," I said. "I'm sure you two will have a great time together."

Olivia stepped toward me and gave me a hug, thanking me before she and Domino left the arena.

CLINICAL NOTES

DR. PETERS

Domino's behavior was not a problem of temperament, motivation, or defiance. It was the predictable outcome of reinforcement delivered without clear contingency. Learning depends on the nervous system being able to link a specific behavior to a specific outcome. When reinforcement is frequent but nonspecific, that link dissolves. Instead of learning what works, the brain learns only that something might work. Anticipation replaces understanding.

From a learning perspective, this falls squarely within the framework of operant conditioning, which describes how animals learn the relationship between their own actions and the consequences that follow. Consequences can involve adding something or removing something, and they can either increase or decrease the likelihood of a behavior occurring again. In ethical horse training, the focus is on reinforcement, not punishment. Reinforcement strengthens behavior; punishment suppresses it, often at significant cost to welfare and learning.

Positive reinforcement strengthens behavior by adding something the horse finds rewarding, such as food. Negative reinforcement strengthens behavior by removing pressure, allowing the horse to return toward comfort. In both cases, learning depends on *precision*: the consequence must follow the intended behavior clearly and consistently.

In Domino's case, positive reinforcement was delivered broadly and indiscriminately. Rewards were offered for nearly any behavior he produced rather than for clearly defined responses. As a result, reinforcement became decoupled from meaning. Instead of strengthening a particular action, it strengthened expectation itself. Domino learned that movement, hesitation, stopping, or shifting might all produce a reward. The striatum—responsible for detecting patterns and predicting outcomes—could not isolate which behavior mattered.

When rewards later became less frequent or stopped altogether, Domino's brain registered a prediction error: the expected outcome did not occur. Prediction error is closely tied to dopaminergic signaling within the striatum. In this case, the mismatch did not produce shutdown or resistance, but hesitation and disruption in forward movement. His nervous system was searching for the contingency that had never been made clear.

Clarity returned only when structure and consistency were reintroduced. Calm, precise cues restored a meaningful cue–response–outcome relationship. Pressure and release once again carried clear information, reducing ambiguity and lowering cognitive load. Importantly, rewards were offered only after Domino's autonomic state showed regulation, allowing reinforcement to support learning rather than drive anticipation.

As the session progressed, movement was shaped through clearer contingencies and increased variability in pattern. This reduced dependence on an external reinforcer and allowed Domino to re-engage with the task itself. Learning stabilized not because reinforcement was removed but because it was finally made informative.

Domino's case illustrates a central principle of learning: reinforcement is powerful only when it is specific. Without clarity, even well-intended rewards can undermine confidence and forwardness. When contingencies are restored, the nervous system no longer has to guess, and learning can proceed with stability and purpose.

It is not the category of reinforcement that determines the behavioral outcome, but the timing, clarity, consistency, and contextual precision with which it is applied.

CHAPTER 10

SKIPPER AND MISTY

The Separation Issue

MARK

Every once in a while, a horse owner will show up to a clinic wanting to work on a specific issue with their horse, but to do that, they also have to bring that horse's buddy or pasture mate into the arena with them. The reason behind this is almost always because either one horse or the other, or both, have a hard time functioning without the other.

This kind of behavior or mindset is usually framed into one of three categories. Buddy sour (where one horse doesn't want to leave another), barn sour (when a horse doesn't want to leave the area of the barn), or herd bound (when a horse doesn't want to leave the company of pasture mates or doesn't want pasture mates to leave them).

When an owner is faced with a horse or horses with any of these issues, often they will either blame themselves: *My horse doesn't like me, trust me, or want to be with me.* Or they blame the horse: *My horse is being bad, belligerent, or trying to get one over on me.*

In reality, the vast majority of this type of behavior can be traced back to a couple relatively simple things: lack of safety and learned behavior.

There are lots of reasons why a horse wouldn't feel safe going with a human and away from a buddy. Poor fitting tack, physical issues, poor handling or riding, lack of clear understanding of foundational principles, such as how to stop, turn, transition or back up, and riding the horse farther away than they were mentally or physically able to tolerate are just a few.

When it comes to learned behavior, one of the biggest culprits is bringing a troubled horse back to its buddy while the troubled horse is in an elevated or frantic state of mind. In doing so, the troubled horse will quickly begin to associate a frantic state of mind with the reward of getting back to its buddy.

This can happen with both the horse that has been taken away, and/or the horse that has been left behind. Here's just a quick example of how something like that can start.

Let's say we take horse A out of the pasture and lead it away from horse B, who is still in the pasture. Horse B starts having trouble, calling for its pasture mate. Eventually the calling gets louder and horse B becomes more animated, running, pacing, bucking, bumping into the fence or gate.

Horse B continues this behavior the entire time horse A is away, lets say thirty minutes. At thirty minutes, the owner brings horse A back and puts it back in the pen with horse B, but horse B is still emotionally elevated. Horse B then calms down as soon as horse A is back in the pen.

Just like that, horse B has learned that in order to get horse A back into the pen it must be in an elevated, worried or frantic state, and that it has to remain in that state for at least a half hour. The next time the owner takes horse A out, horse B goes right back into that troubled state. If horse A doesn't come back within a half hour or so, horse B is likely to escalate its behavior until Horse A is back. By repeating this pattern just a few times we can very effectively teach a horse to go into and stay in a frantic state of mind for hours, days, and even weeks when it's left alone.

***

I was just finishing up with one clinic participant and her horse when Clark, the next rider rode his horse, Skipper, into the arena. Clark and Skipper were followed closely behind by his wife, Madeline, and her horse, Misty, who she was leading. Clark and Skipper went to the far end of the arena and were riding some circles and figure eights. Madeline and Misty went with them, never getting more than about thirty feet away.

When it came Clark's turn to ride with me, Madeline and Misty came with him. Following introductions and a little small talk, I asked Clark what he wanted to work on. He said that he was having trouble getting clean transitions, especially from trot to canter, but that walk to trot wasn't all that great either.

Madeline and Misty stood only a few feet away from Clark and Skipper the entire time I was talking to Clark, and I finally asked why that was. Clark responded by saying they were pasture mates and couldn't be separated. I asked what he meant by that, and he replied that, if they got the two horses too far away from one another, they would both get very upset.

"How far is too far?" I asked.

"Probably from here to there." Clark replied, pointing to about the middle of the arena, not more than sixty feet away.

"Could we see what that looks like?"

"You want me to take him down there?"

"If you don't mind."

"Okay." The tone in his voice told me he didn't think it was a very good idea.

He started riding toward the middle of the area but only got maybe twenty or thirty feet away before Misty started to have trouble, whinnying, pawing, and dancing around Madeline. As soon as she started having trouble, Skipper began slowing, nickering, and trying to turn around.

I asked Clark to start bringing him back but to stop as soon as Misty started to settle down. He got about halfway between where he'd been and where she was when Misty started to calm down. Her head was still high, and she was still whinnying a little, but she was standing still.

I asked Madeline if I could take Misty, and with what looked like a smile of relief, she handed me the lead rope. I then mentioned to Clark that I thought one of the reasons he might be having trouble with transitions was because Skipper and Misty were so attached to one another. He was likely unable to focus on what Clark was asking because he was always so worried about her, where she was and what she was doing.

I suggested we spend some time working on seeing if we could get them to feel better about being separated. He agreed but also told me they'd tried to work on it in the past with both horses but hadn't made much headway.

The goal when working on separating horses as emotionally attached as these two is to replace a worried state of mind with a calm state of mind. I usually do this by finding the more worried of the two and placing that horse in a pen. Then I'll take the other horse away but only to where the horse in the pen first starts having trouble and stopping there. Usually, when the worried horse sees that the other horse isn't going any farther, it starts to calm a bit.

When that happens, we slowly start to close the distance toward the horse in the pen. We would then keep repeating this. As the horse in the pen starts understanding a calm state of mind is what brings the other horse back, we then start increasing the distance. Over time, separating the two is no longer an issue.

This situation was a little trickier, however, because both horses were almost equally worried without the other. Plus, we didn't have a smaller pen available to put one of the horses in while we worked on the process. As a result, I decided to try something a little different.

I asked Clark to ride Skipper in a twenty-foot clockwise circle, while I led Misty in a twenty-foot counterclockwise circle. The two circles would intersect in the middle. By doing this, the horses would be roughly forty feet away from each other when at the far end of their respective circles and nearly able to touch each other where the circles intersected.

At first, they both had a little trouble at the far end of the circles, but after a few laps, they both settled down. Once they settled, we walked them past each other and basically switched directions, so Misty and I were on the clockwise circle and Clark and Skipper were traveling clockwise.

In a pretty short period of time, and repeating this process, we had expanded to thirty-foot circles then forty-foot circles with both horses staying relaxed. Not long after that, we were each doing circles that took up half of the arena without the slightest bit of worry in either horse.

Then I asked Clark to ride Skipper all the way around the arena while I stood with Misty by the arena gate. Neither horse showed any signs of being troubled. By this time, we were getting close to the end of our session and, seeing as how Clark's original concern was his transitions, I went ahead and suggested he ask Skipper for a walk to trot transition.

He was going away from Misty and me at the time of his request, and Skipper responded with a smooth and soft trot without much of a physical aid at all. Misty stood next to me with her head down and eyes half closed. Clark and Skipper did a few more transitions in both directions, and then it was time for his session to be over. He rode to where Misty and I were standing and dismounted. I handed Misty, half asleep, back to Madeline.

Unfortunately, Clark and Skipper were filling in that day due to a cancellation, so it was the only day we got to work together. As a result, we all discussed a game plan to help maintain or improve the work we'd done, which mostly consisted of looking for ways to reward the state of mind we were looking for and, in turn, allowed for a more proactive approach for the horses to build on their own successes.

I was at that venue for another week doing a series of private lessons, and a few days later, Clark stopped by and told me that the change in their two horses had been like night and day. They were both much more settled and while there was still a little worry when they took one horse out of the pen without the other, with just a little work, they were able to help them both settle quickly.

It never ceases to amaze me how seemingly insurmountable issues with our horses can almost always be remedied when they can feel safe through developed understanding. Misty and Skipper are sure nice examples of that.

CLINICAL NOTES

DR. PETERS

Misty and Skipper were expressing a fundamental biological truth about the equine nervous system: while horses are capable of self-regulation, their physiology is often more efficiently stabilized through relationship. Horses are social mammals whose nervous systems continuously reference the bodies, movements, and emotional states of others. A horse can settle alone, but proximity to familiar companions provides an additional layer of safety that can lower arousal more quickly and with less effort. In bonded pairs, this process, known as social buffering, becomes especially pronounced.

When two horses form a strong attachment, their physiology begins to coordinate. Proximity, shared routines, and predictable interaction increase oxytocin signaling, which moderates threat detection within the amygdala and dampens sympathetic activation. Over time, this oxytocin-mediated sense of safety becomes tightly associated with the presence of the other horse. When that presence changes suddenly, the buffering effect weakens. Arousal rises, vigilance increases, and the nervous system shifts into a search state: *Where is my partner? Am I safe?*

Serotonin plays a quieter but equally important role in this process. It contributes to emotional tone, behavioral inhibition, and the ability to pause, settle, and redirect attention. In strongly bonded horses, access to this serotonin-supported settling can become contingent on proximity. When the partner is too distant, the nervous system may struggle to downshift. Attention narrows, scanning increases, and curiosity gives way to vigilance.

In Misty and Skipper's case, each horse's nervous system had learned to reference the other as a primary regulator. Distance, movement, and emotional state were continuously monitored. Even modest separation triggered sympathetic activation: elevated posture, vocalization, tightening through the body, difficulty with transitions, and reduced capacity to process a rider's cues. None of this was willful or emotional in a human sense. It was the predictable output of two nervous systems that had learned safety almost exclusively through each other.

The approach used that day—opposing circles that gradually intersected and expanded—worked not because it forced the horses to "get over" separation but because it allowed their nervous systems to experience controlled fluctuations of distance without crossing into panic. At the far ends of the circles, each horse encountered a brief rise in uncertainty. Near the center, proximity restored the familiar oxytocin-mediated sense of safety. With each lap, this rhythmic alternation created a predictable pattern.

Predictability is one of the brain's strongest antidotes to threat. As the pattern repeated, the nervous system began updating its internal model. Distance no longer reliably predicted loss, danger, or dysregulation. Separation could occur within a window of stability. Over time, serotonin-supported settling became easier to access, and oxytocin's buffering effects began to extend beyond immediate proximity.

What the horses were learning was not endurance, but contingency: calmness, not agitation, was what restored connection. Regulation became the pathway back to safety.

By the time Skipper was asked for transitions away from Misty, his nervous system was no longer fully absorbed in maintaining proximity. Cognitive bandwidth returned. The frontal and motor planning systems, often overshadowed during vigilance, were again available. This is why his transitions softened. His brain was no longer burdened by constant partner tracking. Misty, observing Skipper move calmly at a distance, was able to settle even further. Their physiology shifted from co-reactivity to co-regulation.

What unfolded that day was not simply behavioral training; it was neurobiological restructuring. Both horses learned that separation can occur without loss of safety, and their nervous systems reorganized around that understanding. This is why Clark later found them more settled at home. The learning was not specific to that arena or that exercise. It was the acquisition of a new internal template, one in which regulation no longer depended on constant proximity but could be carried forward into the wider world.

CHAPTER 11:

APRIL

The Horse with Bloodhound Skills

CRISSI

The frayed cotton lead rope in my fist was attached to a nervous buckskin mare named April. I watched as the shiny hindquarters and flagged tails of fifteen horses disappeared down the dusty road. The drums of their hooves faded from my hearing as I gathered the rope in my hand, drawing the mare closer, hoping she didn't have a mind to run after them. Trish, an older woman who had come with the horses that the camp leased for the summer, stood next to me, mouth agape.

I was twenty that year, and the director of a riding program at a Girl Scout camp in northern New Mexico. As a treat at the end of two weeks of lessons, ten girls, me, two wranglers (Debbie and Trish), and a camp counselor would head out into the woods for a trail ride and overnight camping, pitching tents in a field that had a split rail corral for the horses. On this particular outing, Trish said, "We could save hay, hobble the lead horses, and let them graze in the field." After questioning her about the lack of fences, she said, "We do it all the time. The herd will stay together."

I oversaw as each girl untacked and brushed their sweaty horse. The wrangler hobbled her gelding and turned him loose in the field. He put his head down and grazed. One by one, the horses were turned loose, taking large mouthfuls of long summer grass. Trish hobbled Debbie's horse as she held him. April and I stood close by, letting the long hot day pass into memory as we bathed in the stillness of grazing horses. The sun was low in the sky by then, casting pink and orange streaks across the pale blue. I could hear the distant laughter of the girls as they readied for an early dinner and a campfire, complete with s'mores.

What changed? Thirty odd years later, I can guess. But I still don't know. One minute the herd was quiet, the next moment Trish's chestnut gelding lifted his head, neighed, and pitched into a slow lope. Even hobbled, he was moving quickly.

"I thought you said the rest of the herd would stay with the lead horses?" I could see the sweat running down Trish's neck, her blond braid almost as frayed as the rope in my hand. I looked at Debbie, daughter of a New Mexico rancher, as she tried to undo the hobbles from her horse while holding on to the lead rope as he pranced around her.

"They usually do," Trish said. "I don't know what got into them!"

She threw the hobbles for April into the dry grass and went to help Debbie. I walked with April closer to them and said to Trish, "Boost me up!"

She finished unhobbling Debbie's roan gelding before cupping her hands under my bent left knee. I yelped in surprise when she almost threw me over the other side of April. With the lead rope in one hand, and April's long black mane in the other, we wheeled toward the road. I tightened my legs around her barrel, moving with her as she launched into a gallop.

After finding my balance on April as she ran, I realized that riding a galloping horse bareback was far easier than riding a saddled horse in a trot. April let out a high-pitched whinny that knifed through the forest. If she heard anything in response, I couldn't tell. The wind was a low moan in my ears, my sweaty hands gripping the lead rope and her mane.

It was about this time that I realized April's focus was finding the herd. I was, quite literally, along for the ride. I bent lower over her neck, letting her movement pass through my body. I felt the exhilaration of speed. The confidence that only a twenty-year-old could have filled me from my toes that sped above the ground to the top of my sweat-stained cowboy hat. As I relaxed, thinking we were going to gallop all the way back to camp, April tucked her hindquarters and slid to a stop, almost pitching me off over her lowered neck. She began sniffing at a pile of manure. I could feel her ribcage under my legs like bellows. Her eyes were half closed, her ears pricked forward. She lifted her head, flaring her nostrils as she swung her head right and left.

The road we brought the girls down that morning to get to the camp site wasn't the same road the herd had taken. I looked around as April smelled, realizing I didn't recognize this part of the forest. April leaped into another gallop, turning right. I held on.

April stopped and smelled fresh manure twice more. Both times I was ready for her, taking a moment to assess with my eyes while she gathered information with her nose. The last time she stopped, there were two piles of manure. One of them looked older to me, maybe from a horse that morning.

She compared the piles as her breathing slowed. I waited for her to decide which way the herd went. April walked, then she transitioned to a lope, staying to the right. We made two more turns before I realized we were close to being back at camp. I heard a distant whinny, and April sped from a lope to a gallop, neighing in response. As we passed the Girl Scout camp sign, I saw the herd swarming like a school of fish ahead of us, their bright red, black, brown, and gray bodies circling against the closed pipe gate that we had ridden through that morning.

April slowed to a walk, and she touched noses with several of the herd, taking in their breath and sending hers back out. She stopped by the gate. I thought it was time to get off. I looked over my shoulder as Debbie trotted up, swinging her leg over her gelding's hindquarters and dismounting.

"Horses!" she said.

An hour later, we'd herded the horses back to camp and into their corral for the night. The ride back to the barn the next morning was slower than when we came out. For the rest of that summer, there wasn't any mention of hobbles or letting the herd graze.

CLINICAL NOTES

DR. PETERS

April's tracking behavior offers a compelling example of how equine neurobiology supports spatial memory, olfactory navigation, and social cohesion under conditions of sympathetic arousal. The sudden flight of the herd was likely triggered by a subtle environmental cue, perhaps an unfamiliar scent, sound, or movement that activated the sympathetic branch of the autonomic nervous system. In such instances, the locus coeruleus releases norepinephrine, a neuromodulator that heightens arousal, sharpens vigilance, and primes the brain and body for rapid action. While peripheral adrenaline increases heart rate and mobilizes energy, it is norepinephrine in the brain that initiates a state of alert responsiveness. Herd contagion further amplifies this effect, with one horse's arousal rapidly spreading to others through sensory and social feedback loops, a hallmark of prey species dynamics.

Though restrained by a lead rope, April did not remain reactive. Instead, she engaged a network of sensory and cognitive systems to locate her dispersed herd. Her stops to investigate fresh manure piles and scent trails suggest activation of her highly developed olfactory system, including both the main olfactory epithelium and the **vomeronasal organ**, which detects pheromones and other

VOMERONASAL ORGAN (VNO)

Also known as Jacobson's organ, the *vomeronasal organ* (*VNO*) is part of a horse's chemosensory system. Located just above the roof of the mouth, it detects pheromones and other chemical signals related to identity, emotional state, and social interaction. Horses activate this system with a behavior called the flehmen response, curling the upper lip and inhaling deeply to draw scent molecules toward the VNO. It's one of the ways horses gather rich, nonverbal information about their world.

PLACE CELLS

Place cells are specialized neurons found in the hippocampus, a brain region involved in memory and navigation. These cells become active when an animal is in or thinking about a specific location. Together, place cells form an internal map of the environment, allowing animals to remember where they've been and figure out how to get where they're going. In horses, place cells likely work alongside olfactory and visual cues to help them navigate open terrain and remember safe routes.

social chemosignals. These cues can communicate identity, emotional state, and direction of movement. Olfactory information is processed not only as smell, but as meaningful data relayed directly to the amygdala and hippocampus, key regions for emotional relevance and memory encoding. The hippocampus, with its specialized **place cells**, likely enabled April to construct and update a cognitive map of her environment, using scent as a spatial anchor. Her behavior suggests an integration of chemical cues and spatial memory to infer the herd's path, even in unfamiliar terrain.

April's ability to stop, assess, and choose a direction also reflects engagement of the entorhinal cortex, where grid cells estimate direction and distance, complementing the hippocampal spatial framework. Upon reuniting with the herd, her calm demeanor and mutual sniffing exchanges likely reflected a shift from sympathetic arousal to parasympathetic regulation, supported by the release of **oxytocin**. This neuropeptide, often referred to as the "bonding hormone," promotes social affiliation, downregulates fear responses via the amygdala, and helps restore physiological calm.

In sum, April's journey was not merely a feat of instinct but the product of a highly adaptive, integrated nervous system. Her behavior illustrates the nuanced role of norepinephrine in arousal, the olfactory system in social and spatial recognition, the hippocampus and place cells in navigation, and oxytocin in reestablishing emotional balance. What might appear as an act of intuition was in fact a coordinated orchestration of neurobiological systems evolved to promote safety, connection, and the return to calm.

OXYTOCIN

Often called the "bonding hormone," *oxytocin* is a neuropeptide that plays a powerful role in social behavior. In horses, oxytocin is released during calming interactions such as mutual grooming, sniffing, or simply standing close to familiar companions. It helps reduce stress by dampening activity in the amygdala and supporting the parasympathetic nervous system's "rest and digest" functions.

CHAPTER 12

BANJO

A Genetic Gamble

CRISSI

In 2016, two years into healing from a horse accident where I'd been hospitalized with a traumatic brain injury, I admitted that I needed to retire my clinic horse Ally. She struggled physically with being hauled long miles, and it was time for her to slow down and for me to graduate to an unknown horse. I was in the beginning stages of rebuilding my confidence and wanted a horse I could feel safe with but who would also stretch what I knew. I thought about looking for a Missouri Foxtrotter. I'd lost my Foxtrotter, Jack, five years before, and he'd been integral to my growth as a horse person, not to mention one of the best trail horses I've had the honor of riding.

Perusing online ads, doing Google searches, looking locally and nationally, I discovered the market for gaited horses had exploded. Any horse in my budget was a yearling or a horse in their teens who looked as though they needed a good long rest, not more rides, and certainly not thousands of miles in a trailer. After a month of searching, I resigned myself to looking on social media. I joined a page dedicated to gaited horses for sale and there he was; a handsome black gelding stood tied to a tree, ears forward as though he sought to be closer to the person taking the photo.

When I inquired about him, the owner, Anne, said she planned to go to graduate school, and Banjo needed a home where he would be kept active. I laughed.

"My husband and I teach clinics all over the country," I said. "He'll be active enough."

She sent me videos, photos, and answered all my questions. He sounded like a horse that would work for us. He had experience on trails, working cattle, standing tied, and going through water. She had done ranch work with him, and I was especially impressed when she sent a video of her riding him at his gait on a long and loose rein. So many of the Foxtrotters I'd seen at clinics and for sale either paced or were jammed into a bit with spurs and crops. This guy, all 14'3 hands of him, stepped into and out of his gait easily. She sent a video of her riding him bareback at a canter. Another short video showed him Foxtrotting in the field. Anne told me he'd been started under saddle at seven years old, and she got him after he'd been with a trainer for sixty days. He'd been a ranch horse ever since. Banjo was everything I was looking for: he had trail, hauling, and ranching experience; he'd lived in a field for seven years before being ridden, and he was the size I preferred. We completed the details of his sale.

I'd asked if Banjo had papers, and Anne said he did, emailing me a copy. I was further convinced I'd made the right choice when I saw that on his sire's side, he had some of the same bloodlines as my previous Foxtrotter, Jack. When I looked at the bloodlines on his dam's side, a sliver of doubt embedded itself in my certainty.

His dam's parents were father and daughter. Meaning, the same horse that sired the female (Banjo's grandmother), was mated back to her to produce Banjo's dam.

Inbreeding isn't a rare practice, and it's not a new concept. It isn't restricted to certain breeds, and its kissing cousin, line breeding, has also been used to produce horses who are thought to be genetically superior. Line breeding can be understood this way: breeders use line breeding to strengthen desirable traits, believing it reduces the risks of inbreeding. The aim is to pass on a superior ancestor's qualities without overly limiting genetic diversity. Inbreeding, from what I've talked with breeders about, is used to further increase the odds of concentrating those desirable traits.

That's one side of the inbreeding/linebreeding story. The other side is what I know from my time spent with horses all over the world: linebreeding and inbreeding are a gamble. The foal is born, and maybe the conditions are right for him or her to develop normally, performing at high athletic levels. It's just as likely that the foal will be born, and no matter how much repetition it's put through, it cannot make it past being haltered. Or saddled. Or ridden, even at a walk. We've seen defensive inbred horses, especially from the Premarin industry. We've seen inbred Mustangs; many times, the social hierarchy of the herd is permanently disrupted during gathers, and sisters and brothers and mothers and sons end up mating. There's no way to track who is breeding to whom. We've seen inbred Warmbloods, Thoroughbreds, Arabians, Morgans, and Quarter Horses.

When I looked at Banjo's papers and compared them with the videos, nothing changed my mind. I asked Anne what she thought of his inbreeding, and she said, "I saw that too when I bought him, but it didn't bother me. He's happy to work, and I've had zero problems with him."

I slept on this information, and upon waking the next morning, I called a shipper and scheduled to have him hauled from Missouri to Colorado.

A week later, Banjo was in our barn, spending his time in the run outside instead of the stall attached to it. He ate and drank well, and besides being nervous about his new surroundings, he seemed to be everything I'd seen in the videos. Anne was right: he was a fundamentally happy horse. He was curious, engaged with people, and made me laugh with his water trough antics.

We took it slowly with him. We hauled to clinics nearby, and for the first month, I would saddle him but not ride. He stood tied and stayed calm in clinic situations where horses are coming and going, and the hours are long. He had a little trouble with trailer loading, but by also taking this slowly, he improved.

Two years went by, and I was riding Banjo in clinics at both a walk and his Foxtrot. He could stand tied, had been hauled across the country numerous times, and ate and slept well, whether he was in Florida, Colorado, or North Jenina. He still wasn't a big fan of stalls when that was the only option to over-night in, but if his buddy, Top, was next to him, it seemed he could tolerate it, once he was able to cross the threshold. Changes in the type of footing or the light and dark contrast seemed to trouble him. He'd cock his head sideways, snort, and raise and lower his head. I stood inside the stall, waiting for him to follow, which he did after being given time to sort it out. I chalked this up to how his brain worked and was okay with it.

On a cold February morning, at our stop in Santa Fe at a friend's ranch, I first witnessed the limitations of inbreeding. Mark and I got up early so we could get to our Arizona venue before dark. This was when the days are short and the weather unpredictable; every minute counts, and our schedule on travel days tends to be about using our hauling time wisely. We don't want our horses in the trailer any longer than they have to be.

Being the smaller of our two horses, we loaded Banjo first. I could see my breath in the dawn's light. Mark held his horse, Top, while I did what I'd done for years: I stepped up into the trailer and asked Banjo to follow. He pulled back, the whites of his eyes showing. I waited until he released the pressure on the lead rope and asked him in again. He bolted backward, rearing when he hit the end of the rope, taking me with him for several feet. He stood still, every muscle in his body tight, steam rising from his body. It was most curious how a horse could go from knowing how to load to behaving as though he'd never been in a trailer.

Mark stepped in to help, but another thirty minutes passed before we convinced Banjo to load. It was thirty minutes of snorting, trying to bolt, and rearing to get away. This wasn't fighting; he was trying to escape. He was terrified.

After arriving in Arizona that evening, we unloaded him from the trailer and then asked him back in. He snorted and put his nose on the floor, then he loaded himself in slow motion. Each evening when the clinic finished, we repeated this process until he relaxed going in and coming out. That was seven years ago, and he's been loading calmly since.

***

Our summer clinics last ten days with a couple of five-day clinics sprinkled in. During those ten days, we worked cattle, rode with people in a large field, and worked in an arena. Banjo wasn't bothered by horses close to him; he stood tied for hours, and he seemed to enjoy working cattle, though he wasn't very responsive to my cues if I needed him to move sideways. Since we'd come so far in his education as a clinic horse, I thought I needed to spend more time on our lateral work: turn on the haunches, turn on the forehand, and sidepass.

By the end of the summer of 2017, Banjo understood to turn on the haunches and turn on the forehand, but sidepassing was difficult. As a gaited horse, all four of his limbs can move in distinct patterns, so I wasn't too worried about him one day understanding how to move them sideways. *There's always next summer*, I thought.

The next summer, though, when I asked for a turn on the forehand, Banjo shifted his body and feet, but more out of confusion than knowing what to do. We reviewed how to do the movement, along with making sure he knew how to give to pressure from the bit. He'd understood it the previous fall and had all winter on pasture to integrate the information.

At the end of ten days, we'd made some headway, but Banjo's understanding wasn't where I'd left him the summer before. Backing up was sticky, giving to the bit was sticky, and his lateral work was hit and miss. All that summer, and all those rides, I worked on the same concepts. I was determined to be patient and take my time, giving Banjo consistent exposure to the skills I'd like him to have.

In between times, Mark and I would trail ride, and this is where he excelled. Anne had done a great job with him. Banjo would lead, ears forward and body relaxed. He was sure-footed and not prone to spooking (except on the trails in Florida, when the Saw Palmetto would rattle as the deer jumped out of them).

The next summer played out much the same—with no discernible progress in his understanding. I decided to give lateral work, backing, and softening to the bit a rest and see how he felt about cantering. When I asked, I felt his whole body tighten as his head came up.

We went to the round pen, and when I asked him to canter without me on his back, the same thing happened: his whole body tightened, his head came up, and he bounced around the pen like a deer.

By the end of the summer of 2019, he cantered comfortably with a saddle, without me on his back. I tried again in 2020.

Aren't we fortunate to have a passion that requires us to be outside? When the 2020 pandemic hit and our out-of-state clinics cancelled, we could resume our summer schedule at Happy Dog Ranch with modifications. This allowed me to keep getting to know Banjo.

That summer, I again worked with Banjo on the same skills, and again, the information either didn't stick, or it scared him. I began to feel a lot of doubt about my own skills, so for one clinic, I left him at home and brought Ally out of retirement. I asked Ally for lateral work, softening to the bit, and backing, and it was as though only days had gone by, not years. When we brought Banjo to the next clinic to work, nothing had changed, except my doubt and frustration had increased.

After work one evening, I admitted to myself that the frustration came from my stubborn belief that with enough time and repetition, Banjo could learn, even though I knew that a primary hallmark of an inbred horse is their inability to retain information. I'd been in denial about Banjo's lack of capacity to add anything to what he already knew. His coordination issues showed up more frequently, which was also weighing heavy on my mind. I gave up on asking him to canter while I rode him, figuring if it was that stressful, with no apparent change despite all the help I could give, we might as well let it go. I could feel my self-doubt easing as I faced Banjo's reality.

After a long, difficult look at where we were, and after chatting with Mark, I decided to accept everything Banjo was able to do, and do well, and stop trying to train him to do or be anything else. We spent another sunny summer together, focusing on his strengths. We stopped hauling him to clinics.

These days, Banjo is in his mid-teens, and he and one of our granddaughters are fast friends. She rides him bareback around our corral, brushes him, hangs out with him, feeds him apples, and plays with his long black mane. I couldn't have asked for a better girl to hang out and learn with a good horse. Banjo continues to bring laughter to my days, and I'm now grateful for who he is and what he taught me.

CLINICAL NOTES

DR. PETERS

Banjo's story illustrates how neurobiological and genetic factors can shape a horse's learning profile, behavioral responses, and adaptability over time. From the beginning, Banjo demonstrated many strengths such as emotional stability, a positive affective tone, strong social engagement, and solid trail performance. These traits reflect healthy function across several brain systems, particularly those governing basic autonomic regulation, **sensory integration**, and affiliative behavior. But the difficulties that emerged around skill retention, motor coordination, and response to pressure suggest a different side of the neurological picture, one that may be linked to his inbreeding history.

Inbreeding increases the probability of homozygosity, where genetic material is inherited in duplicate from closely related ancestors. While this can reinforce desirable traits, it can also unmask deleterious recessive alleles, genes that may disrupt the development of the nervous system, neuromuscular coordination, or cognitive processing. In horses, as in other species, inbreeding has been associated with reduced neuroplasticity, learning challenges, and increased prevalence of neurological anomalies. These outcomes may not be

SENSORY INTEGRATION

The horse's brain is highly attuned to subtle changes in the environment. Sensory input from sight, touch, balance, and body position is constantly being processed by the nervous system to create a sense of safety and alertness. In some horses, especially those with neurological sensitivity, this *sensory integration* doesn't always run smoothly. A small change in footing, a shift in light and shadow, or an unfamiliar sound can trigger hesitation or panic. Offering time, patience, and predictability helps these horses build confidence through repeated safe experiences.

INBREEDING

Inbreeding occurs when closely related horses are bred, increasing the likelihood of inheriting two identical copies of a gene. While some breeders aim to "lock in" desirable traits, inbreeding can also expose hidden genetic risks. These may include issues with brain development, coordination, memory, and emotional regulation. Horses affected by inbreeding may have difficulty learning new tasks, retaining training, or coping with changes in environment.

visible early in life but can surface when demands for cognitive flexibility, memory consolidation, or complex motor planning increase.

Banjo's struggle with retaining learned skills, such as lateral work, cantering under saddle, and consistent softening to pressure, suggests potential disruptions in procedural learning circuits, particularly those involving the cerebellum (coordination), hippocampus (memory), cingulate gyrus and frontal cortex (adaptability and decision-making). His tendency to startle at footing changes or contrast boundaries, such as the threshold between light and dark in stall entries, may reflect difficulties in sensory integration or pattern recognition, functions often modulated by the parietal cortex and vestibular system. These subtle sensory processing challenges can amplify anxiety or result in exaggerated motor responses when a horse encounters novel or ambiguous stimuli.

The traumatic loading episode in Santa Fe offers a glimpse into how a combination of sensory hypersensitivity, context-specific fear memory, and impaired cognitive inhibition can converge. The initial bolt and rearing, followed by a gradual but successful reconditioning process, shows both the imprint of past trauma and the capacity for recovery when supported by patience and consistency. This speaks to the adaptive role of the amygdala in threat detection and emotional memory, and the potential for extinction learning through repeated safe exposure, a process that, while slower in some horses, can still lead to rewiring when approached with care.

That Banjo thrived in low-pressure, familiar contexts such as trail rides, relaxed social interactions, and time with children, speaks to the enduring strength of his limbic system, particularly in generating trust, positive affect, and social regulation. The oxytocinergic system, central to social bonding and emotional resilience, appears to be intact and well-expressed in his relationships, especially in his bond with the granddaughter. His tendency to seek connection, explore novel stimuli like water troughs, and maintain strong affiliative behaviors despite cognitive challenges is not only endearing but biologically significant.

Ultimately, Banjo's story is a case study in neurological asymmetry: certain systems in his brain are robust and adaptive, while others show signs of constraint or fragility. His case underscores the importance of seeing each horse as an individual, one whose learning style, emotional capacity, and behavioral repertoire are shaped not only by training and environment but also by the architecture of their nervous system. Recognizing these boundaries is not a failure of training, but an act of empathy grounded in neuroscience.

By honoring what Banjo could do, rather than perseverating on what he could not, Crissi was able to model a kind of relational neuroplasticity, changing expectations and behavior in response to the horse's nervous system, not in spite of it. This kind of adjustment reflects the best of what a horse-human bond can become: a co-regulated partnership built, not just on performance, but on respect for the brain behind the behavior.

CHAPTER 13:

CASSIE

The Troubled Bay Mare

MARK

This isn't good, I thought to myself.

The bay mare had been a hot mess the day before when the owner rode her into the arena. No sooner had they come in than she began jumping, spinning, shaking her head, trying to rear, offering to bolt, and bucking, sometimes all at the same time.

The owner explained, in between airs above the ground and careening back and forth across the arena, that Cassie was an off-the-track Thoroughbred he bought as a rehab project a few years before. The behavior we were seeing, while a little worse than normal, was not at all uncommon. He also said she got regular bodywork, her teeth were done by a natural balance dentist, he spent a fortune getting a saddle that fit her, and he had tried any number of bits and bitless bridles, none of which seemed any better than any other.

The owner was very experienced and was able to skillfully stay with the anxious mare without too much trouble. As a result, we were able to spend most of the next hour working on improving his timing on both redirecting her outbursts and also releasing when she found a way to relax, if even for only a second or two. The better the owner's timing got, the longer Cassie could go in a more relaxed state. By the end of our time together, the little mare had settled down enough that she was able to quietly and seemingly thoughtfully walk about halfway around the arena.

The owner told me he had been able to get her that quiet in the past but not until she wore herself to a frazzle. This had been the first time she quieted so quickly. I was optimistic that what we had done that day made a difference for her and told the owner that hopefully she would feel at least as good tomorrow when we start as when we finished that day. With any luck, she might even feel a little better.

Unfortunately, that wasn't to be the case. This day not only was she not any better, but she actually seemed worse. We went back to what we had success with the day before, only this time it didn't really matter. It was as if Cassie not only hadn't retained any of the information she received the day before, but what we were doing was making her even more anxious, if that was possible.

And that wasn't good.

That's when I asked the owner the question that I probably should have asked the day before: Does it feel like you're starting over every time you work with her? His answer was yes. I then asked if the mare was registered, and she was. I mentioned that we may want to look at her papers.

Back in the 1980s, I began noticing unusual behaviors in some horses that made working with them very difficult and in some cases downright impossible. Initially, there were two behaviors that stood out: the horse not being able to retain information from one day, week, or month to the next, and aggression that began intermittently and slowly increased in severity and frequency. A few years after first noticing these behaviors, I happened to read an article on inbreeding in dogs, and interestingly, some of the negative side effects of inbreeding in dogs were the very same behaviors I was seeing in horses.

From that point forward, when presented with these behaviors in a horse, I would ask the owner if the horse was registered, and after checking the papers in those that were, found pretty much all had either inbreeding or linebreeding in their lineage.

Over the next forty or so years, we were able to directly associate other unwanted behaviors to registered horses who had been either inbred or line-bred. These were intermittent bucking, bolting, and the slow and sometimes sudden onset of spooking or even panicking when presented with familiar things or situations.

We also have noted certain birth defects, such as deformed palates and underdeveloped mandibles, as well as degenerative bone diseases that could also be attributed to inbreeding or linebreeding. Along with that, we were also able to correlate the onset of certain coordination issues to inbreeding, not the least of which is the horse's progressive unwillingness to transition to faster gaits, both while alone in a pasture and while being ridden.

Most recently, we have found evidence that suggests in some cases, the negative effects of inbreeding can skip as many as eight generations before randomly showing up again. Today, there is even technology to test for inbreeding in horses who aren't registered. We know of at least one case in which a mustang who presented some of the negative behaviors associated with inbreeding had been tested and was found to have been as little as 13% inbred. And yet the behaviors were still there.

Unfortunately, it's been my experience that many owners and trainers who run across the types of issues we're talking about don't know to investigate inbreeding or linebreeding as a possible cause for the unwanted behavior. As a result, folks often mistake the horse's behavior as adversarial in nature and, to get the horse to stop the behavior, end up upping the level of force being used to get the horse to comply. Seldom, if ever, does doing so ever end well for anybody, especially the horse.

Still in other cases when the inbred horse's behavior becomes too unmanageable for the horse to be usable in any other way, some breeders resort to turning the affected horse back into breeding stock, effectively perpetuating the same issues repeatedly.

But it's not all bad news. In fact, here's the really interesting part: from what we've seen, horses who end up afflicted by the negative effects of inbreeding appear to be somewhat random. In other words, not all inbred horses show the negative behaviors we've mentioned. In fact, the effects (or lack thereof) that inbreeding and linebreeding have on the horse population seem to be as wide and varied as the horse population itself!

On one end of the spectrum, there are inbred horses with seemingly no adverse effects whatsoever, including everyday using ranch horses and world champion show horses. On the other end of the spectrum are horses with behavior so dangerous that they have had to be put down. In the middle are horses whose behavior might be referred to as odd or quirky (but harmless and manageable) to horses who have had to be retired or used strictly as a companion horse.

The horse I mentioned earlier was lucky enough to become one of these. After that second session with Cassie, the owner pulled up her papers online and found she was, in fact, inbred. Her lines showed her sire and his full sister were her parents, and her dam was the product of her sire and his mother.

After finding this out, the owner gave the mare away as a pasture mate where she lived out the rest of her days in the quiet company of a gelding who had no companion. Seeing as how most negative effects of inbreeding only show up when the horse is asked to do anything other than just be a horse, there could have been no better outcome for the little bay mare.

CLINICAL NOTES

DR. PETERS

The behavioral profile of Cassie, marked by persistent hyperarousal, lack of learning retention, and disorganized movement, raises strong concerns about underlying neurodevelopmental dysfunction, likely compounded by inbreeding. Although skilled handling temporarily calmed her, the absence of carryover from one session to the next suggests a breakdown in the brain's capacity for experience-dependent plasticity. In healthy horses, repeated positive exposures strengthen synaptic connections through a process known as long-term potentiation (LTP), particularly in the **hippocampus** and **cerebellum**. This enables memory consolidation and the integration of learned motor patterns. In this case, the complete loss of gains made from one day to the next points to either a failure in forming those connections or a rapid degradation of them due to chronic stress or structural limitations.

HIPPOCAMPUS

The *hippocampus* is a key structure in the brain's limbic system, responsible for forming and consolidating memories, processing spatial orientation, and interpreting context. In horses, a well-functioning hippocampus allows the animal to link past experiences with present situations, essential for learning, recognizing safety cues, and adapting to changing environments. When hippocampal function is compromised, a horse may struggle to retain training across sessions or may react to familiar settings as if they're entirely new. This disruption can make even simple routines feel unpredictable or unsafe, undermining confidence and consistency in learning.

Unlike Banjo, whose temperament remained steady despite some cognitive challenges, the bay mare presented a more globally unstable profile, marked by volatility, poor sensory integration, and minimal ability to downregulate **arousal**. Where Banjo retained social engagement and showed strong affiliative bonding, suggesting intact oxytocinergic function and limbic stability, this mare struggled to maintain emotional or behavioral continuity even in low-pressure settings. The contrast illustrates how inbreeding can produce a spectrum of neurological effects: in some horses, only specific circuits (like motor learning or spatial memory) are affected, while in others, core systems governing arousal regulation, memory, and behavioral inhibition may all be compromised.

Such patterns are well documented in the scientific literature. Inbreeding has been associated with reduced hippocampal volume, impaired learning and memory, cerebellar abnormalities affecting coordination, and heightened baseline sympathetic tone. Studies in both equines and other mammals, including dogs, cattle, and rodents, have shown that inbreeding increases the likelihood of homozygosity at **deleterious loci**, which can affect the development of neural networks responsible for emotional regulation, sensory integration, and cognitive flexibility. These disruptions often surface not early in life, but under pressure when learning demands increase or when environmental challenges require adaptive regulation.

In this mare's case, the inability to maintain calm or integrate prior learning is consistent with impairments in the prefrontal-limbic system, where context appraisal, inhibition, and arousal modulation take place. Her heightened reactivity to familiar settings and stimuli may reflect sensory processing dysfunction, in which benign environmental input is misclassified as threatening due to poor neurosensory gating. This is particularly common in horses with cerebellar or parietal dysfunction, often expressed as spooking, bolting, or panic in response to subtle changes in footing, light, or spatial contrast.

Ultimately, while Cassie showed moments of engagement and brief success under saddle, her nervous system appeared unable to stabilize or rewire with experience. Her genetic history revealing close inbreeding between full siblings

CEREBELLUM

Traditionally known for its role in balance and motor coordination, the *cerebellum* also contributes to rhythm, timing, and even aspects of emotional regulation. In the horse, the cerebellum helps fine-tune gaits, manage transitions, and produce fluid, adaptive movement. A horse with cerebellar inefficiency may appear clumsy, uncoordinated, or delayed in responding to cues, particularly during gait transitions or spatial navigation. These subtle motor deficits can sometimes be misread as resistance, when in fact they reflect real neurological limitations in processing and synchronizing movement.

AROUSAL STATES

Arousal is the nervous system's activation level, ranging from deep calm to high alert. Healthy horses move up and down this range as the day changes. When regulation is lost, a horse may get stuck at either extreme. The *arousal states* include hyperarousal and hypoarousal. Hyperarousal (often driven by high norepinephrine) can look like spooking, bolting, reactivity, or an inability to settle—always "on watch"—while hypoarousal can look like shutdown or learned helplessness; the horse seems quiet but is disengaged rather than relaxed.

13

and parent-offspring likely introduced or amplified the neurodevelopmental vulnerabilities at play. In such cases, when repeated training fails to take hold and arousal cannot be downregulated, it is not a failure of the handler but rather a limit set by neurobiology itself. Offering the mare a life free of demand, in a quiet social setting, was not a resignation but a compassionate choice, an adaptation to what her nervous system could and could not do. It serves as a reminder that not all horses are equipped for human use and that recognizing those limits is an essential part of ethical horsemanship.

DELETERIOUS LOCI

A locus is simply a gene's "address" on a chromosome. With inbreeding, a horse is more likely to inherit two copies of the same gene. That can strengthen good traits, but it can also bring out harmful hidden genes called *deleterious loci.*In some cases, these genes can affect how the brain and nervous system develop or function. The horse may seem normal early on but later show learning difficulty, poor coordination, or unusual fear and reactivity. Because the signs can look like a training issue, it's important to remember that sometimes the limit is biological, not behavioral.

REHABILITATION:

Am I Safe?

CHAPTER 14

MOUSE

The Troubled Roping Horse

MARK

I've always had a soft spot for those older horses who've been good at their jobs but are maybe starting to get to the end of their careers. I suppose that was why I was drawn to Mouse, a seventeen-year-old semi-retired roping horse.

The first time I saw Mouse was at a clinic venue in the high country east of San Diego in the fall of 2001. He wasn't in the clinic, but rather I saw him standing in a pen near the arena in which the clinic was held. He had recently been purchased by the clinic host's husband, Jim, who was an experienced rider but was getting into roping and needed an experienced rope horse to help him develop his skills.

The gelding was mouse-gray in color, thus his name, and I remember thinking that he looked tired that first time I saw him. During the four days of the clinic I would occasionally see Jim riding Mouse. Sometimes he'd be sitting on him while tossing his rope at a plastic roping dummy, other times they'd be moving cattle around in the pen adjacent to the arena, while still other times they would just be cruising around the property.

Nothing overly concerning stuck out for me during any of those times that I saw the two of them riding together, which was why it was a bit of a surprise one morning to see Mouse in a panic, literally trying to climb out of his pen. From what I could see, the panic seemed to be due to Jim going in the pen while carrying a halter.

During a passing conversation with Jim later that day, I would learn about Mouse's history. Born in Montana, he'd been a ranch horse who had been passed around from owner to owner and state to state until he finally ended up at a ranch in the San Diego area. He'd been at that ranch for several years, and according to Jim, the owner there took great pride in making sure his horses were always afraid of him.

When it came to Mouse, he had succeeded in spades. Mouse was next to impossible to catch, which is why he was in the small pen. But even in the small pen, things could get pretty dicey, like they had earlier that morning. In a big pen, he would need to be roped in order to catch him, if you could get close enough to him to do so.

Grooming was always a challenge, and saddling or cinching him was a disaster most of the time. He would stand tied all day, until or unless someone would approach him. If they did, Mouse would pull back with everything he had, and if he broke away, good luck catching him.

He didn't like to be petted; he was head shy, particularly around his left ear. He wasn't too fond of having his feet handled or being bridled and could be a bit of a challenge to get on. Once on, he was solid most of the time, unless someone approached him on the ground. On a good day, if someone came up to him, he would shy away. On a bad day, he would bolt, but he would never just stand still. To his credit, he was solid when roping or working cattle, but, according to his previous owner, spurs and a bit with enough leverage to get him stopped once he got going was a must.

Jim told me he knew Mouse had issues when he bought him but didn't really know the full extent. He got him mostly because he was cheap and had the basic skills he needed to get him started on learning how to rope. He planned on keeping him, at least for a while, but would probably move him along once he found another horse with fewer issues and that he could afford.

The next time I saw Mouse was in the spring of the following year when I returned to the same venue to teach another clinic. Apparently, Jim had been trying to sell Mouse, but his reputation as a troubled horse always preceded him, so there had been little or no interest in him. Jim told me he didn't feel like Mouse was any worse than when he got him, but he wasn't any better either.

As I said at the outset, I've always had a soft spot for older working horses in the later years of their life, and although I wasn't really in the market for another horse, I told Jim I'd take him.

***

To be honest, I didn't really know what I would do with Mouse once I got him home. I hoped, at a minimum, I could give him a place where he could live out his days in peace. At best, maybe I could find a way to help him feel better than he had been.

Once home, I turned him in with our herd with the idea that he could just be a horse for a while without anybody asking anything of him. We had twelve head at the time—three mares that we'd raised and nine geldings that we had acquired over time. The herd itself was emotionally balanced, and all got along well together. Other than about fifteen minutes of initial pushing, chasing, and squealing when Mouse first went into the herd, things quieted down and went back to normal relatively quickly.

At first, I noticed that anytime I went in the pen with the horses, Mouse would get as far away as he could while the others would usually come and check out what I was doing. After about a month, Mouse would keep a safe distance but was no longer going to the far end of the pen. Not long after that, he would stay with the horses when they would come around to see what I was doing but still stayed a safe distance from me.

This went on for quite a while until, one day, nearly three months after he had arrived, I was kneeling in the pen repairing a small section of fence when I felt the light brushing of a horse's muzzle on my back. I didn't think much of it, assuming it was just one of the horses from our established herd. Then there was another soft nudge on my back, and still another. I glanced over my shoulder to see it was Mouse.

My turning surprised him, and he snorted, quickly backing away about four or five feet. I was pretty much done with what I had been doing, but instead of getting up and leaving, I turned back to the fence. Several minutes passed before I felt another soft nudge on my back, then another, and another. I heard a deep sigh, and then the slow sound of Mouse walking away.

Over the next several weeks, and after a few more times of Mouse approaching me when my back was to him, I decided to see how he'd feel with me approaching him. Now, I think when people hear me say I would work on approaching him, they might think that I was trying to get up to or even pet him. That wasn't it at all. I just wanted to see how he'd feel if I moved toward him but not necessarily get up to him.

I learned a lot in a short period of time about the very fine line he had between me moving just a hair too fast or being just a hair too close when I offered to move toward him. It was as if he had a very specific boundary line that I was welcome to come up to but not cross. Initially, that line was about forty-five feet away. As soon as I moved even the tiniest bit past that line, he would be gone.

The next couple months between the two of us were painstakingly slow, but in that time, Mouse taught me more about communication and conversation between horse and human than I ever thought was possible. Six months had passed since I brought him home before he let me walk up to him, pet him, and slip a halter over his nose without worry.

But that was just the first step. As I mentioned earlier, the list of things Mouse was not okay with was much longer than the things he was okay with. Saddling was such a major issue for him that, if not done just right every time, he would panic and blow backwards with everything he had. While he did feel marginally better over time, saddling remained a worry for him until the day he died.

While he was okay with me catching and handling him, and eventually he would be okay with Crissi catching and handling him, he was very leery of anybody else. He was the same way about his feet being handled. Crissi and I could handle them, as well as our farrier, who he seemed to really like, but nobody else.

Early on, I came to understand that the problem he had with bridling stemmed from some severe damage he had done to his tongue. This became evident one day when I saw him yawn, which exposed a big, deep scar across the entire width of his tongue. I rode him in a hackamore, and the bridling issue went away, as did most of his head shyness.

By the time a year had passed, Mouse was doing so well that he had become my main clinic horse and basically the horse I went to whenever I needed to get any job done. He had become so responsive during that time that all I had to do was think about something and he'd do it. This can be seen briefly in the documentary *The Path of the Horse*, in which Mouse and I are working cattle together without me giving him any physical aids.

To give you an idea of his responsiveness, I remember once while doing a clinic in Florida, I was sitting on Mouse near the arena fence having just finished up with one student and waiting for the next. It was a very warm day and, having been in the same spot for several minutes, Mouse had cocked a foot and closed his eyes. I'd let the rein just drape over his neck. It was about then that an auditor came up to the fence.

"Why do you do that?" she asked, pointing to Mouse.

"Do what?"

"Why do you let him go to sleep like that? It'll take forever to wake him up so you can get a job done."

I reached down and gently touched the rein with my index finger. Mouse squared his stance, slowly raised his head, and glanced back at me. After a couple seconds, I took my finger off the rein, Mouse cocked his hind foot and lowered his head.

The woman smiled. "Never mind," she said.

While Mouse eventually felt a little better about people walking up to him when I was on him, he never did get to feeling any better about people he didn't know walking up to him, especially when he was tied. At every clinic, I would always mention over the loudspeaker that people should not approach Mouse when he was tied. If I needed to tie him while working with a horse and rider from the ground, I would always tie him somewhere in the arena where I thought nobody could get to him.

Still, it would seem some people just couldn't help themselves, and inevitably, someone would go up and try to pet him, which always resulted in him panicking and pulling back.

"That's why I ask that people don't approach him," I would say with varying degrees of sternness in my voice.

Their response would almost always be something along the lines of, "He looked so quiet. I thought it would be okay."

***

Mouse was with us for the last seven years of his life. During those years, we never regarded him as a horse that needed pity for what he'd been through, or as one that needed rescuing from a bad situation, or one who needed rehabilitation from unfortunate handling. We simply saw him as part of our herd and treated him as such. Like all of our horses, we wanted to help him build on the things he did well, develop understanding where understanding was missing, and help him manage those things that were difficult, if not impossible, for him to do.

In return, Mouse was ultimately able to show us that, even though he may have once been down, he was never really out.

CLINICAL NOTES

DR. PETERS

Mouse's behavior reflects what happens when a nervous system spends years in environments where human interaction consistently predicts threat. In such horses, fear circuitry does not simply activate in response to the present moment; it becomes biased by history. The amygdala and its associated networks learn to treat approach, handling, and confinement as predictors of danger. The horse is no longer responding to what is but to what the nervous system has learned to expect. The past becomes the template through which every new interaction is filtered.

In chronic stress, the bed nucleus of the stria terminalis (BNST) plays a particularly important role. While the amygdala generates rapid, short-lived responses to immediate threats, the BNST sustains longer-duration states of vigilance. When handling has been unpredictable, forceful, or fear-based, the BNST can remain chronically engaged, maintaining a background state of alertness even when no obvious danger is present. The horse lives in anticipation rather than reaction.

Mouse's difficulty being approached, haltered, saddled, or led while tied fits this pattern precisely. His nervous system had learned that safety depended on staying ahead of threat, monitoring proximity, movement, and restraint. In this state, even neutral or gentle handling can trigger defensive preparation because the system is already primed.

Importantly, recent neuroscience helps explain how change becomes possible in these cases. Research has identified a population of dopamine-responsive neurons within the amygdala that contribute to fear extinction, not by signaling reward, but by firing when an expected aversive event does *not* occur. Each time Mouse anticipated a threat and it failed to materialize, this signal helped update the brain's prediction. Fear was not erased, but its certainty began to weaken.

Serotonin plays a complementary role in this process. By moderating amygdala output and supporting emotional tone, serotonin helps the nervous system tolerate uncertainty long enough for new information to register. As defensive circuitry gradually softened, these repeated "nothing bad happened" signals created the conditions for change.

Only after this shift did the mesolimbic dopamine system, the brain's motivation and approach network, begin to re-engage. This pathway, originating in the ventral tegmental area and projecting to the ventral striatum, nucleus accumbens, amygdala, and frontal regions, does not generate pleasure itself. Instead, it encodes prediction and expectation: *Is this worth approaching? Is effort likely to lead to a beneficial outcome?*

As Mouse experienced consistent handling, clear boundaries, and reliable release, dopamine signaling began reinforcing approach rather than avoidance. Curiosity replaced scanning. Engagement replaced withdrawal. This shift did not happen because he was "braver" but because his nervous system was finally receiving coherent information.

Crucially, this change did not occur through desensitization or **habituation**. It occurred through predictability, time, and control, and, most importantly, through the consistent availability of safety. Safety is not merely the absence of threat. It is the presence of conditions that allow the horse to retreat, observe, and re-engage on their own terms. Mouse's boundaries were not pushed or challenged. They were respected. That respect prevented overwhelm and allowed the nervous system to reorganize without being flooded.

SEROTONIN

Serotonin is a neurotransmitter that helps regulate a horse's emotional balance and ability to pause and recover after stress. When serotonin systems are working well, horses tend to be more calmly attentive, less impulsive, and better able to shift from vigilance back toward safety, supporting learning and connection.

Serotonin also affects the body, especially the gut and sleep. Because much of the body's serotonin is linked to the digestive system, changes in diet, pain, routine, or social stress can influence how a horse feels and behaves.

HABITUATION

Habituation is the brain's way of learning that a repeated, non-threatening stimulus can be safely ignored. With gentle, predictable exposure, neural activity in sensory and emotional circuits decreases, allowing the horse to adapt, conserve energy and remain calm.

Some early learning, however, remained. His defensiveness when tied, his uneasiness with unfamiliar handlers, and his sensitivity to certain forms of touch never fully disappeared. This does not represent failure. When vigilance has been maintained for years, some pathways become deeply consolidated. Neuroplasticity allows softening, stabilization, and improved flexibility, but it does not erase history.

Progress in cases like Mouse's is not measured by the elimination of all reactivity. It is measured by increased ease, reduced baseline vigilance, improved recovery, and greater agency. Mouse's story illustrates both truths of the nervous system: that chronic stress leaves enduring neurobiological traces, and that meaningful change is possible when a horse is given clarity, consistency, and the felt experience of safety. The goal is not to undo the past but to help the horse live well with the nervous system they have now.

CHAPTER 15

ZEPHYR

Life Out of the Fast Lane

CRISSI

"I'm thinking of buying a horse," my friend Jen said. She'd called me earlier that day, wondering if I could come over and look at him with her. It was 2005, and by that time, I had two other geldings and had been training horses for ten years.

I looked at the round pen where she was keeping him. I saw a leggy bay gelding, covered in sweat as he trotted and cantered along one side of the pen. His head was raised high, and he was scanning the horizon, as though he could see his way back home. By the way he moved, I could tell he was athletic and well put together.

"What did he used to do?" I asked.

"He's been an endurance horse for the past ten years. The owner wanted to find a good home for him."

Jen went on to tell me that he was fifteen and had been raised and trained by a local woman I knew who put quality time into her horses and usually didn't start them under saddle until after they were three. The gelding's name was Zip, and his sire was an Arabian stallion crossed with his dam, who was half Trakehner and half Thoroughbred.

"Well, I can see why they used him for endurance."

Jen nodded. "I'm not sure I'm going to take him, though."

"Why not?" I asked.

She shrugged. "He might be too tall."

I laughed and said, "He's at least 16 hands."

Jen said, "16'2. All my horses aren't much more than fifteen hands."

At the time, I had two horses and a barn with four stalls with runs. A third horse didn't seem out of the question, so I asked Jen to let me know what she decided because I might take him if she didn't want him.

Two days later, she called and said she wasn't going to take Zip. I drove over to Jen's to pick him up and give his owner a check. I held a large brown envelope that had his papers, health history, and receipt, watching the dust follow the previous owner's car as she drove away.

The high-headed gelding was easy to load in my horse trailer. He pawed the floor the whole way across town. Once home, leading him to a pen, one that was in a different paddock than my other two horses, was like hanging on to the end of a kite; he pranced and threw his head, snorting and pulling against the lead rope. I got him into his pen with a pile of grass hay, a salt block, and a couple of buckets of water. I watched as he ignored the food and water, whinnying to my other two horses who stood on their side of the fence, hind legs cocked and ears forward.

By the end of the day, hunger and thirst had kicked in, and Zip was grabbing bites of food in between pacing the fence. Looking at him, I swore he'd dropped weight in the six hours he'd been at home.

When a horse is sold and taken to a new place, it's disorienting. Nothing is the same—the smells, the water, the hay, the people, the horizon, even how the light looks at certain times of the day. Sometimes I wonder if it feels to them like they've landed on an alien planet, and they don't know if they'll survive or not.

At the time, I knew that the two weeks that he'd live in the separate paddock was a good time for him to settle in. I thought he could get used to me being around, feeding and mucking, and he'd start to feel better. I thought once he was in with my other two geldings, he'd relax even more.

We did get used to each other. During those two weeks, I decided to rename him Zephyr. I liked the idea of him being a gentle breeze, rather than a term for speed. He'd stopped pacing as much, but I noticed that he didn't lie down like the other horses. I hadn't even seen him roll. I could halter and groom him, and the only saddle in my tack room that fit his narrow back was my dressage saddle. I'd taken him out for walks down the road I lived on, as well as did some ground work in my small arena. He was still alert, his head high and his body tense, but he'd not gone back to prancing or pulling on the lead rope. All in all, I was looking forward to seeing how he felt about being ridden.

Although he was tense while being groomed and tacked up, he didn't seem upset by it. I bridled him and led him to the arena. He had a little trouble at the stump I used for a mounting block, so we spent some time helping him understand how to stand still. It took about forty minutes, but by the end of it, he was lowering his head, blowing, licking his lips and looking sleepy. The sun was setting by this time, so I called it a day. Even though I didn't ride like I'd planned, it seemed Zephyr had done a lot of work.

The next day at the mounting block, he was quieter. After I slipped my right foot through the stirrup, he took off at a trot. It was only four of his long strides before he reached the fence, turning abruptly. His head was high, and I was posting higher than I ever had before; my stirrups were too short. I turned him to the left, toward the center, and he slowed for a stride before rushing forward again. I turned him right, and he slowed, then rushed forward. When I put some pressure on the bit, he was able to walk for a couple of rushed strides before breaking into a trot again. We went along this way, me adjusting my balance with my too-short stirrups, him eating up the distance in my little arena with his long strides.

I noticed that if I asked him for a large figure eight, we could slow down at certain points in the turns. I started to add a little pressure through the reins at those places, seeing if we could slow down to a walk. After twenty minutes of this, we walked. Even though his head was still high, he maintained his walk, snorting and blowing. I kicked my feet out of the stirrups and settled into the saddle. As we made another turn, I added some pressure to the reins, and he stopped. We walked and stopped several more times before I dismounted. Once I was off, he lowered his head and shook his mane.

This was often the pattern with Zephyr in the arena for the next several months. We were making progress, but it was slow progress. I knew that his previous job required speed and endurance. I knew he'd often been ridden quickly over a lot of miles. I suspected that him having a certain job wasn't the only thing feeding his worry. He still wasn't laying down, and even after being ridden and hosed off, he didn't roll. I called my vet.

After her visit, and her recommending we treat him for ulcers, I felt along his back and noticed he had some sore spots. The next visit was from an equine bodyworker I'd been using for a couple of years. After that, a veterinarian who only did hand floating of horses' teeth. After that, specialized shoes from my farrier. During this time, I was reading about equine nutrition and ways to help horses gain weight without adding sugar. My herd already got free choice hay, but then I learned about soaking beet pulp and grass pellets. To this, I added some herbs and magnesium for gut support, ground flax seeds, and rolled oats. I was doing more cooking and meal planning for my horses than I was for myself.

Zephyr physically thrived, so much so I'd see him standing in the sun, dozing with my other two geldings. If I took one of them away to ride though, he went back to running the fence, whinnying, and sweating. Any weight we'd added was soon gone.

One of the central turning points for Zephyr and I was asking Mark for his advice. I laid out everything I'd done, all the ways I'd tried to help, and my frustration that Zephyr still couldn't find a way to calm down. Mark listened, nodded, and said, "I noticed that, when you lead him, he's pretty close to you. Do you want him there when he's upset?"

"Not really. I can't seem to find a way to get him to walk quietly, though. He's always nervous."

We spent the next hour showing Zephyr where the boundaries were while leading. By the end of the session, I could lead him anywhere, and he understood to stay an arm's length away. The next day, Mark showed me how Zephyr wasn't breathing, so we longed him to help him begin to take deeper breaths. The third day was another leading and longeing session, and by the fourth day, I was riding at a walk and practicing the halt, which was teaching Zephyr what giving to pressure meant.

He was fifteen years old, and while he'd done his endurance job well, there were things missing in his understanding. As a young trainer, I'd missed that his lack of understanding of the skills I was asking for was, in fact, part of his anxiety.

I was on a mission: every time Zephyr and I spent time together, we practiced leading with boundaries. This consistency from me allowed him to calm down. I only had to longe him twice more before he could breathe more efficiently. When we rode in the arena, I stayed at a walk, asking him to lower his head and give to pressure. When I took my other geldings out to ride, they only went back into the main paddock when Zephyr was calm. It was the first time I'd understood that working with a horse's state of mind was far more potent than training them to do stuff. The stuff showed up, and more beautifully, if the horse was in a calm state of mind first.

The next year, I began working with Zephyr on the trail. At first, I'd tack him up and lead him over my favorite trails in the cinder hills of northern Arizona. We'd go down the road about five hundred feet from the barn and turn back around. It became a fun experiment for me, to listen to Zephyr and turn around before he got worried. After we could get a mile down the road to the trail head, I began riding him, repeating the pattern. It was only a month before we went on a trail ride.

The best day of our time together came when we were heading down the trail toward home one evening. It was September, and the warmth of the fall day was melding into the sunset. Zephyr lowered his head to the ground, I thought to smell something, but then he walked that way for several strides. When he brought his head up again, it was at the same level of his withers, his neck relaxed and long, his stride loose. He sighed. We rode the rest of the way home like this, long and relaxed, with me doing everything I could to hold in joyous hoots that Zephyr, who once thought that riding meant speed, could now walk calm and relaxed.

I spent that year hauling Zephyr to different places and took him to see Mark again for five days. My goal was to help Zephyr feel calm in a new place. By day five, Zephyr and I were once again able to ride out on the road and across the field at a loose and easy walk. He loaded calmly to go home, and when I put him back in the paddock with his two herd mates, he bent his front knees, cocked his tail to the right, and lowered himself to the ground, grunting as he rolled.

***

Lizzy was a student of mine, a nice rider who'd been taking lessons with one of my other geldings. She often commented on what a nice horse Zephyr was. I always agreed. One day, she asked if I'd ever consider selling him to her.

I had to think about it for a while. Zephyr and I had learned so much from each other, and our time on the trail had evolved into something I looked forward to, and he seemed to also enjoy. I hadn't seen him pacing or being nervous in many months. He was gaining weight, and often I'd come home from work and see him lying down sleeping, the other horses standing near him. If he went to a new home with a new person, would he still be able to keep the worry at bay?

The next time Lizzy came over, I asked if she'd like a lesson with Zephyr. I figured if anyone would know if it would work, he would. She groomed and tacked him up, and I showed her how to lead him and what we were looking for at the mounting block. Through all of this, and her riding time, he was calm, his head low, and he didn't get nervous even while she learned a couple of new skills.

Lizzy kept riding Zephyr, and they worked great together. We started going out on the trail, her on Zephyr and I on one of my other geldings. I got to see what his loose and long walk looked like from a different perspective. His eyes were alert and his ears forward, but everything in his body was swinging; tail to ribcage to nose, he was relaxed.

I was going to be moving out of state soon, so after Lizzy's lesson, I handed her the same brown envelope I'd received with Zephyr's history in it.

"He's yours, if you'd still like to take him home."

She barely got out the word "Yes!" before wrapping me in a huge hug.

I saw Lizzy and Zephyr a couple of times before I moved. He was calm and relaxed, and Lizzy adored him. Before I lost touch with Lizzy, last I heard, he was doing great, and they were having fun.

There were so many lessons I learned during our time together. Thousands of horses later, I think of Zephyr often, especially when we have a horse who is nervous, and an owner who doesn't know if their horse will (or can) calm down. The lesson that still sticks with me, and what I hope to share with people, is that so much more is possible when we learn how to carry the calm inside of ourselves.

CLINICAL NOTES

DR. PETERS

Zephyr's story is a clear example of how behavior, brain function, and learning are inseparable. What looks like "training" on the surface is often the visible expression of a nervous system that is either regulated enough to learn or too activated to process information cleanly. In Zephyr's case, the early signs—high head carriage, constant scanning, reduced appetite, and difficulty resting—pointed to a horse living in persistent high arousal.

When a horse stays keyed up for long periods, the body's stress system tends to stay active. The brain signals the adrenal glands to release stress hormones, including cortisol, to mobilize energy and keep the animal ready for action. In the short term, this can be useful. But when stress becomes chronic or predictable, especially when human interaction repeatedly predicts pressure, performance, or speed, cortisol can remain elevated more often than it should. Over time, this can interfere with digestion, disrupt sleep and recovery, and make the horse more reactive and harder to settle. In horses, prolonged stress also correlates with gastrointestinal vulnerability (including ulcer risk), appetite changes, and reduced capacity for calm social engagement.

Zephyr's reluctance to lie down or roll, even in a safe environment, was one of the most telling signs. Lying down requires a deep internal sense of security because it temporarily reduces the ability to respond quickly. When a horse cannot access that state, it often means the nervous system is still interpreting the world as potentially unsafe. If the body is braced for danger, it prioritizes vigilance over rest. This state is typically accompanied by higher levels of arousal chemistry such as norepinephrine (the "alertness" signal) and reduced access to the calming systems that allow the horse to soften and recover.

This is also where the neuroendocrine system matters. The neuroendocrine system is the brain's way of controlling the body through hormones. In simple terms: the brain does not just think—it signals. It signals the heart, the gut, the immune system, muscle tone, and motivation. When the hypothalamus and related stress pathways stay activated, the horse may look "wired" or "guarded," but the deeper truth is physiological: the body is running a program designed for survival, not learning.

Zephyr's early behavior was therefore not "bad behavior." It was a predictable output of an incomplete learning history combined with unresolved stress. An endurance background can build remarkable fitness and grit, but it doesn't always provide the kind of foundational experiences that build cognitive flexibility, impulse control, and clarity with pressure-and-release cues. Without that clarity, ambiguity itself can feel threatening. And when ambiguity feels threatening, the horse shifts from being responsive to becoming reactive.

A key piece of this story is interoceptive awareness, the brain's ability to read the body from the inside—breathing, heart rate, gut discomfort, pain, temperature, and internal tension. Horses live through interoception constantly. A sudden refusal, spook, or change in "attitude" can be the brain responding to internal discomfort or rising arousal, not willful disobedience. When Zephyr's system began to settle, his internal signals likely became less urgent and less aversive, allowing him to make different choices.

What facilitated Zephyr's transformation was not simply time. It was intentional neuroplasticity, guided experiences that allowed his brain to form new associations and build new routines. Neuroplasticity is the brain's capacity to change based on repeated experience. It is biological: neural connections adjust according to what is practiced, what is reinforced, and what becomes predictable.

Crissi's approach succeeded because it provided predictability, agency, and clear cause-and-effect, three elements the nervous system requires for durable change. The slow groundwork reduced uncertainty through repeated, patterned interactions. It gave Zephyr the ability to pause, observe, and re-engage without being trapped. And it restored clarity through calm boundaries and consistent signals the horse could understand.

The structured groundwork, such as leading at an arm's length, systematic longeing to promote deeper respiration, and reinforcing calm responses, did more than "teach manners." It created a reliable rhythm that Zephyr's brain could model. The basal ganglia, the brain's habit and pattern system, thrives on repeatable sequences. When those sequences are calm and coherent, the nervous system begins to treat them as safe. Over time, new procedural memory forms: not just what to do, but how to feel while doing it.

There is also an important learning principle at work: the brain updates rapidly when outcomes are better than expected. When a horse anticipates pressure or discomfort but instead experiences relief, safety, or a clear release, the nervous system registers that mismatch and revises its prediction. In plain language, the horse expected one thing, but reality proved safer. Those moments when surprise resolves into safety can accelerate learning because they mark the experience as meaningful.

Crissi's most pivotal shift was moving from task-oriented riding goals to steady attention to Zephyr's internal state. That is not "being soft." It is being precise. A regulated handler can stabilize a dysregulated horse. This is co-regulation: one nervous system helping another return to balance through predictable cues, good timing, and the absence of escalation. Over time, this kind of interaction supports recovery physiology; the horse can breathe more fully, soften posture, digest, rest, and eventually begin to trust.

As Zephyr's internal state improved, the external signs followed: slower breathing, lowered head carriage, improved appetite and rest, and later, greater confidence on the trail. His willingness to lie down and rest was not merely a behavioral milestone; it was neurobiological evidence that his system had begun to experience the environment as safe enough to release vigilance.

Neuroscience reminds us that learning and healing are not separate processes. They are built into the same machinery: experience, prediction, body state, and relationship. Zephyr's trajectory shows that even deeply conditioned autonomic patterns can be reshaped, not through force, but through safe, repeatable interactions that give the horse clarity and control.

For both horse and human, growth is not created simply by the absence of fear. It is created by the presence of calm, carried consistently enough that the nervous system can finally believe it.

CHAPTER 16

AMIR

The Misunderstood Trail Horse

MARK

From the very beginning, the clinics I did while on the road had always been one on one. That is, I would work with one horse and rider at a time for about an hour. When their time was up, and/or we had gotten the pair to a better place than when they came in, they would be done, and another rider would come in. From time to time, there might have been a little overlap between riders when a second rider would come in while I was still working with the first in order to warm their horse up or just get them used to the arena, auditors, and P.A. speaker.

Normally, this wasn't an issue; we would seldom use the entire arena during our sessions anyway. As a result, the second rider could come in and use one end of the arena while I continued to work with the first rider on the other end. But on this day, things were a little different.

I was about fifteen minutes from finishing up with a young woman on a nice warmblood who had been working on lead changes when the next rider attempted to come in. I say attempted because the nearly sixteen hand Arab gelding, saddled and bridled, head high, and screaming at the top of his lungs, was crashing into the fence, the gate, and into his owner as she frantically worked to get the heavy pipe gate to the arena open.

A friend of the owner rushed to her aid and swung the gate open in time for the owner and her horse to burst in. Things weren't much better once they were inside. With the owner on foot, the horse raced circles around her as she clung to the reins and did everything she could to keep from being run over.

I have to admit, I was a little surprised and more than a bit concerned when I saw the owner trying to get the gelding over to the mounting block in what looked like was going to be an attempt to get on. I suggested to her, over the loudspeaker, instead of mounting up, to take the horse to the nearby round pen and turn him loose inside so he could move without being restrained. The owner's friend opened the gate, and the pair stampeded out.

A few minutes later, after finishing up with the last rider, I made my way over to the round pen. Not surprisingly, the gelding was still racing around the inside with his owner, Carly, standing outside, bridle hanging over her shoulder.

After brief introductions, I was able to find out from Carly that she bought the gelding, whose name was Amir, four months earlier. He was eight years old, coming nine, and a former champion halter horse that she had hoped to turn into a trail horse.

I asked if Amir, along with his halter training, had also been trained to ride. She told me no, but that her brother Dan, who was a roper, had gotten on him and ridden him out on the trail.

"How'd that go? I asked.

"Amir was pretty bratty, like today."

"Is that the bit your brother used?" I pointed to the bridle on her shoulder, to which was attached a fairly heavy leverage bit with a chain curb strap.

"Yeah, but Dan said it would be what I needed to get him to pay attention."

I then asked if she knew how Amir was kept in his previous life: in a stall, stall with a run, a paddock, pasture.

"He'd always lived in a stall," she said. "I turned him in with my other gelding when I got him home, but they didn't get along, so now I keep him in his own pen."

"Near the other horse?"

"On the other side of the barn."

Carly told me that it took several weeks for Amir to settle down after she separated the two geldings, but now he seemed pretty okay as long as he was in his pen. He would "get bratty" when she took him out to try to work with him.

"He's almost nine years old, and his previous owner hauled him all over the country," she said. "So why does he act like he's never seen anything?" It was an honest and valid question.

I started by explaining that, even though Amir was nearly nine and had been on the show circuit for most of that time, his actual life experience was very limited. For instance, Amir's life was more than likely limited to going from his stall into an arena, then maybe a trip to the wash rack, and then back into a stall with the occasional trailer ride to a show sprinkled in.

As a halter horse, he probably didn't get turned out much, and if he did, it was probably by himself so he wouldn't be bitten or kicked by another horse. That was probably at least one of the reasons he wasn't able to get along with her other gelding when she put them together. Having never been around other horses, he didn't really know what the rules were or how to act when he was around another horse.

Along those same lines, having spent his entire life in a stall is probably why it took so long for him to get comfortable in his outdoor pen and why he gets worried when she takes him out of it. After all, now that he's feeling safe in his pen, anything outside of it may still feel foreign and unsafe for him, especially if he's being asked to do things he doesn't understand. And that could be anything outside of the one job he's been doing for the past eight years.

"But my brother took him on a trail ride," she said. "How could he have done that if he didn't understand it?"

"Just because a horse does something doesn't mean they understand it," I shrugged. "Plus, there are folks out there that can get a horse to do just about anything...once."

"So you don't think Amir can be a trail horse?"

"I think it'll take some time," I said. "But he's here for the next three days, so let's see how things feel after that."

I felt one of the most important things we would need to do with Amir was to see if we could turn some of his fear into curiosity. Initially I did that by stepping in the pen with him and just giving him some subtle direction as he ran around the inside of the pen. Most of that was just interrupting his patterns and direction of travel with relatively small movements of my own.

I wanted to make sure my movements, which consisted primarily of the orientation of my body in relation to his—quietly raising a hand or arm, stepping toward or away from him, and the like—were big enough to get his attention but not big enough to add to his worry.

Initially, he could find a way to stop and check out what I was doing but wasn't able to stay before going back to the rail and resuming his running. But before long, his being able to stop for a second or two turned into three or four seconds, and that turned into six or seven seconds, and so on, until he was able to stand quietly, head down and body relaxed, for several minutes. From there, we worked on some leading and boundaries with him, which he picked up pretty quickly with both Carly and me, and that's where we stopped.

The goal for that first day had been to just help him calm down enough so he could take in some relatively simple information. To that end, we wanted to find something that was easy enough for him to accomplish but difficult enough that he had to engage with the situation in order to find the answer. Once he was able to do that, we quit for the day.

The pen he was staying in during the clinic was pretty close to the round pen, which I believe helped him maintain some composure when he transitioned from one pen to the other. Carly asked Crissi to do some bodywork on Amir that afternoon, and Crissi said they were able to get a lot of nice work done and he was a perfect gentleman the whole time. In fact, after fifteen minutes of bodywork, he lay down and napped.

The next day, we started where we left off, and Amir did great. So we added lunging at a walk and trot and responding to subtle movement and energy of the handler for transitions and halts. From there, we added ground driving with two lines attached to his halter. In particular, we worked on moving forward, stopping, turning, and backing, all with soft cues and at the walk and trot.

On the third day, we added ground driving with the bit and saddle while also going over everything we did the day before. Carly did the majority of the work, and the two of them did great together.

By the end of that third day, Carly mentioned that she could now see how unfair it had been for her to think Amir could go directly from the show world he'd been in all his life to automatically being a trail horse. She saw firsthand how he simply had little or no understanding of how to do some of the most basic things a trail horse—or any riding horse for that matter—really needed to know. She also saw what it was going to take to help fill in those blanks for him.

She then told me she was going to keep working with Amir toward hopefully helping him be a nice trail horse, but in the meantime, she would find a seasoned trail horse she could ride. That way, she could take as much time with Amir as he needed without feeling like she needed to rush him.

For me, Carly is the epitome of a good horse owner: one that does the best they can and, when they know better, they do better.

Can't ask for more than that.

CLINICAL NOTES

DR. PETERS

Amir's behavior in the arena was not a character flaw, and it was not "brat-tiness." It was a nervous system doing exactly what it was built to do when it believes safety is at risk: escalate, move fast, and try to regain control through distance and motion. The high head, screaming, crashing into the gate and fence, and running circles around Carly were classic signs of an acute sympathetic surge with high arousal chemistry driving flight behavior, with very little capacity left for thinking, learning, or softness.

Carly's question was honest: Amir had traveled the country as a show horse, so why did he act like he had "never seen anything"? The answer is that travel is not the same as life experience. Amir likely lived inside a narrow loop: stall to arena to wash rack to stall, with trailer rides that transported him between versions of the same routine. Even a horse that has been to many shows may have had very little true exposure to open environments, varied footing, unstructured spaces, or calm exploration. When a nervous system grows up inside a tight, repetitive world, it can become highly specialized for that world and easily overwhelmed outside of it.

This matters because early and repeated experiences shape the brain's stress and learning circuits. Limited turnout, minimal social contact, and heavy reliance on restraint can leave a horse under-practiced in the skills we later expect from a trail horse: emotional flexibility, curiosity, impulse control, and recovery after arousal. Amir's difficulty with Carly's other gelding also fits this pattern. Social competence is learned. Horses raised with little herd exposure often don't develop smooth "rules of engagement," so social interaction itself becomes stressful.

In the arena, the primary biological driver was a sensitized threat system. When the amygdala and related fear circuitry have a long history of pairing human handling with pressure, restraint, or performance demand, the brain begins to predict threat before anything "bad" actually happens. The horse is not only responding to the present moment; he is responding to a forecast built from his past. In that state, added control, like a severe leverage bit and curb chain, can increase arousal even if it temporarily suppresses behavior because the nervous system experiences more pressure with no increase in understanding.

Mark's first decision was neurologically wise: remove Amir from a tight, escalating situation and put him in a space where movement could occur without a fight over restraint. The round pen became a place to change the internal state before trying to teach new tasks. This is not permissiveness. It is sequencing. You cannot install learning on top of panic.

The first day was essentially about transforming fear into curiosity. Mark did not "chase Amir until he got tired." He used small, calibrated interventions, body orientation, a lifted hand, a step in or out, big enough to interrupt Amir's looping pattern but not so big that they added threat. This is the sweet spot where the nervous system can begin to pause and ask, *What is happening?* At first, Amir could only stop for a second or two. Then three or four. Then six or seven. This is a measurable sign of regulation: the nervous system gradually shifting from continuous flight into brief windows where attention and processing return.

Those small windows matter because they are where neuroplasticity actually happens. The brain changes when an old prediction fails to come true. Amir predicted escalation. Instead, he experienced interruption followed by neutrality, then relief. The nervous system begins to update: Stopping doesn't make it worse. Looking doesn't make it worse. Being near a human doesn't necessarily mean pressure. Over repetitions, those updates accumulate into a new baseline.

Once Amir could stand quietly with his head down and body softer, Mark moved to boundaries and leading. That sequence is important: regulation first, then simple, clear cause-and-effect. Leading is not a trivial skill; it is a conversation about personal space, predictability, and trust. Amir learned quickly because, now that his arousal had dropped, his brain could actually take in information.

Two details in Mark's account are especially telling. First, Amir's pen during the clinic was close to the round pen, which reduced the stress of transitions.

Many horses can perform inside one "safe pocket" but unravel during the walk between places. Keeping the geography predictable supported learning. Second, Crissi's bodywork led to Amir lying down and napping after about fifteen minutes. That is not merely "relaxation." Lying down is a high-threshold behavior. It suggests Amir's nervous system finally accessed a deep enough sense of safety to shift into recovery mode with parasympathetic physiology returning, digestion and rest becoming possible, and the brain gaining access to consolidation.

Day two added low-arousal movement with structure: lunging at walk and trot with subtle transitions and halts. This is where the nervous system learns to regulate within motion, not just in stillness. From there, ground driving with two lines on the halter built steering, forward, stop, turn, and back—foundational trail skills taught in a way that preserved clarity and prevented overwhelm.

Day three layered in the bit and saddle during ground driving while reviewing everything from the day before. This is another crucial principle: introduce new equipment and sensations inside a pattern the horse already understands. When the brain can predict the sequence, novelty becomes tolerable. Carly doing most of the work also mattered. Amir wasn't merely learning "Mark." He was learning a new relationship with Carly with new timing, new predictability, and new safety.

By the end, Carly's insight was the real transformation: she recognized it had been unfair to expect an instant jump from a stall-based show life to a trail life that demands environmental flexibility, emotional resilience, and basic riding education. Her decision to ride a seasoned trail horse while taking as much time as Amir needed was not only humane...it was neurologically sound. Rushing is one of the fastest ways to re-sensitize a nervous system. Time, structure, and repeated calm success are how you build a trail horse brain.

Amir's story is a reminder that behavior is often biology expressing history. When we change the conditions—predictability, pressure, clarity, and recovery—we give the nervous system a chance to reorganize. Amir did not need to be "shown who's boss." He needed a pathway from panic into understanding, repeated enough times that curiosity could become his new habit.

DANTE

A Metabolically Stressed Horse

CRISSI

Mark and I were working at a barn in Southern California, outside of San Diego. The barn was tucked in between two hills, and they'd designed and used their space in some clever ways, given that there were easily thirty horses on what looked to be about sixty acres. There was an indoor arena and a steel barn that had twenty stalls connected to it by an aisleway that had two wash stalls on one side. There were outdoor paddocks with shelters, an outdoor jumping arena, and two hot walkers, plus a covered round pen.

We were working in the arena that sunny March morning. An older man leading a prancing bay gelding came toward us. I saw Mark working with a woman riding a tall gray horse, warming up over some low jumps. I turned my attention back to the gelding and his owner.

"Hi, I'm Crissi." I smiled, keeping my hands in my jacket pockets because it looked like this guy had plenty to do without trying to figure out how to shake my hand. He gave me a quick smile before asking his horse to lope a small circle around him. I stepped back.

"I'm Greg," he said, switching hands on the long lead rope as the bay spun and went the other direction.

"How can I help you today, Greg?"

He laughed, turning his head toward me. "You're seeing it. I'd like to know how to get him to stand still for two seconds!"

"What's your horse's name?"

"Dante," Greg said, feeding out some rope so Dante could cover more ground.

"Do you have a longe line?" I asked. Greg nodded, pointing a finger to his left.

"The tack room in the big white stock trailer."

"I'll be right back."

After we'd found a second to help Dante stop so we could switch to the longe line, I asked Greg if he'd like me to longe him or if he was okay doing it.

"Nah, I'm good. This is what we do every time I take him out."

"Why don't you tell me a little bit about Dante?"

I made my way into the middle of the large circle where Greg was standing so we didn't have to raise our voices near a horse who had enough going on as it was. We both kept our eyes on Dante as he cantered first one way, and then the other, Greg handling the rope with ease. The difference between his calm demeanor and Dante's energetic one was the first thing I noticed. This was not a nervous owner, and in fact, Greg could easily tell me about his gelding as we walked in a small circle in the middle of the now trotting horse.

Greg told me Dante was an off-the-track Thoroughbred who had gone through a year of rehabilitation in Kentucky before being shipped to California. He'd had three owners in two years, and Greg was the third. Dante was six years old, and Greg had bought him the year before.

"Do you know why he'd been sold so often?" I asked.

Greg laughed, "Dante is a lot of horse."

When I looked at the sweaty gelding walking at the end of the longe line, I agreed. Now that he wasn't so tense, I could see that he was carrying a bit of extra weight and his neck was starting to get cresty. He was shod on all four feet.

"How is he with his feet?"

"Good. The farrier said the last time he was here that Dante seemed a little sensitive, so I added a hoof and joint supplement to his feed," Greg said. "I have to be sure to give him his turnout time before the farrier shows up though. Otherwise Dante isn't too happy to stand still."

I asked Greg what Dante ate in a day. Greg let Dante come to a stop at the end of the line, then turned toward me.

"He gets a flake of alfalfa three times a day, grass hay in a hay net all day, and sweet feed with his supplements."

"Okay, what are the supplements you give him?"

Greg listed six different supplements then paused.

"I should probably also tell you that I use the sweet feed to catch him when his turnout hour is up."

"Okay. Did you want to look at catching at some point?"

"Maybe. Right now, I'm more concerned with how hot he is and what I can do to get him to calm down."

When I asked Greg where Dante lived, Greg told me that he had a stall at the other end of the barn.

"Does it have a run attached?" I half suspected it didn't, but waited for Greg to say that it was a 10 by 12 stall, and Dante got an hour of turnout once a day.

"How often do you get out to work with him, Greg?"

"I live about an hour away, so it's only on the weekends and maybe once during the week, on my day off."

I watched as Dante trotted at the end of the line, his head down.

"It doesn't take him very long to get worn out, but I'd like some ideas on how we can start out this way."

"By any chance has your vet looked at him?"

"She has. Why?"

"Have you talked about insulin resistance?"

"She took some blood, and the results were higher than she liked, but he wasn't insulin resistant. She's the one who suggested cutting back on his alfalfa and adding grass hay."

"Did you notice any change after that?"

"He dropped too much weight, so I put him on sweet feed and added in alfalfa, but less of it. It's what we fed our skinny horses when I was a kid."

"Did you grow up on a ranch?"

Greg said, "More a farm, really. Our horses were working horses, and on their days off, us kids rode them."

I paused, debating how to talk about what can often be a prickly topic for many horse owners. As I watched Greg and Dante, I decided the direct approach was best.

"Greg, I'm not sure Dante's energy is completely a training issue."

"Okay." He asked Dante to stop, and the gelding did, lowering his head.

"What I'm thinking is that for a horse with his activity level, he may be eating too much."

"The last owner said he was a hard keeper, and sweet feed was the best way she found to maintain his weight."

"That's a common opinion. The trouble is, when a horse has too much sugar or starch and not enough activity, it builds up in their system. When we add being in a stall for the length of time that Dante is, it's not unusual for this behavior to show up."

Greg was quiet as he gathered the longe line in one hand. He patted the horse on his neck, seeming to be lost in thought.

17

DANTE: A METABOLICALLY STRESSED HORSE

"You know," he began, then sighed. "I'm not much of a fan of stalls, but at the time I got Dante, this was the only place I could find. He used to get only alfalfa, until the vet told the barn manager that we needed to cut back."

"That's a good start." I'd hoped that, by encouraging him, we could talk more about how to help Dante feel better. "I saw some paddocks out back that have shelters. Are those available?"

"They are," Greg said. "And they're certainly cheaper!" I laughed with him. "The barn manager says it's easier if all the horses are in one place, so she wants us to keep most of the horses in stalls."

"Would you be willing to do a little experiment?" He nodded, leading Dante closer.

I chatted with Greg about putting together a mix that was high in fiber, with as few ingredients as possible, to begin the process of replacing the sweet feed. I told him that a vet I'd known had said that feeding vitamin B1 helps with the process of breaking down the carbohydrates in alfalfa and, for most horses, taking away the common effect of too much energy. I asked Greg if he'd ever talked to his vet about ulcers.

"Yes, and one of the supplements he gets has magnesium in it. It was also supposed to help him be calmer." He laughed, shaking his head.

I nodded. "You might talk with her about other ways she'd recommend supporting his gut and see if that helps him keep some weight on too."

We finished our session with Greg riding Dante at a walk and helping Dante to relax. Once Greg dismounted, he slipped the reins over Dante's head and strode toward me.

"I think I'll go get Butch, Dante's buddy, and take them both to that open paddock before I go to the feed store with the grocery list you gave me."

We agreed that we'd begin by cutting out one flake of alfalfa per day and transitioning Dante from sweet feed to the high fiber mix. Greg would also get the B1 supplement while the alfalfa was still in Dante's diet.

The next morning, as Greg led an already saddled Dante in, I could tell the gelding was feeling better. He was still alert, his head high and his body tense, but he wasn't prancing. Greg had the longe line attached to the leather halter. As he asked Dante to circle him, the gelding first trotted a couple of laps before bolting into a canter, head high. After five minutes, he slowed back down to a trot for several laps, before once again walking. I made my way toward Greg, my boots sinking into the damp sand.

"How'd it go last night?" I asked.

"He was a little nervous in that paddock, even with Butch there, but by this morning, he was quieter. He even ate the new feed I bought him. And I took out his alfalfa at noon."

"How's Dante feeling to you this morning?"

"He was quiet enough for me to groom him and tack up, so that's an improvement."

Over the next two days, Greg and Dante made a lot of progress. Dante's head began to lower as his body relaxed, and we could ask him into a trot and a canter, and he could maintain the speed Greg wanted. By the fourth day, and our last session together, Dante didn't need to be longed at all.

After dismounting at the end of our session, Greg said, "I think I'm going to talk with the barn manager about Dante and Butch living in that paddock. He's acting like the horse I'd hoped he'd be."

"I think that's a great idea."

"You want to know a secret?" he smiled.

"Always!"

"I have a little land where I live. My wife and I have talked about building a barn and paddock for our two horses. After our time together during this clinic, and the changes Dante has made, I think that's the next step."

Six months later, I got an email from Greg with a photo attached. He was standing beside Dante in a paddock, with a red barn in the background. Dante looked relaxed, and like he'd lost some weight. Greg wrote that he took Dante on long walks around the neighborhood and had started trail riding him, with his wife riding Butch. Dante was eating only grass hay and the low starch mix he and I had talked about. The geldings could choose whether to seek shelter in the barn or stay out in the paddock. Greg put a smiley face after saying, "I haven't had to longe him once. His weight is good, he's settling in, and he's really a quiet guy."

I smiled the whole time I was replying to his email, congratulating him on a job well done and commenting on how much better Dante looked. As I sent it off, I leaned back in my chair, reflecting that when we bring horses into our lives, it's beneficial to keep in mind that, one, they are horses, and two, horses have different needs than humans. I still have a lot of admiration for Greg, for putting the well-being of his horses first and working toward keeping them in an environment where they thrived.

CLINICAL NOTES

DR. PETERS

Dante's story illustrates more than a behavioral concern, it reflects a biological imbalance with roots in contemporary horse-keeping practices. Horses evolved to move continuously, forage throughout the day, and regulate their metabolism through physical activity and fiber-rich diets. When this natural pattern is disrupted by confinement, high-starch feeds, and limited social and sensory input, it can dysregulate the entire system.

Metabolic Syndrome in horses, as in humans, is characterized by insulin resistance (IR), obesity, and altered glucose metabolism. Neuroscientifically, this state affects not only peripheral physiology but also central brain processes. Elevated insulin and glucose levels interact with the hypothalamic–pituitary–adrenal (HPA) axis and disrupt normal **homeostatic signaling**, contributing to heightened sympathetic arousal and decreased parasympathetic tone. Horses like Dante, often labeled as "hot" or "too much horse," may be neurologically overstimulated by internal metabolic signals, not just environmental triggers.

Obesity and IR impair insulin signaling in the hypothalamus, which plays a critical role in energy balance, motivation, and stress regulation. Chronic exposure to high-sugar feeds and limited movement can also increase **neuroinflammation** and affect dopaminergic tone, potentially reducing

HOMEOSTATIC SIGNALING

Homeostasis is the body's way of staying in balance by keeping temperature, hydration, energy, hormones, and heart function within healthy ranges. *Homeostatic signaling* is the constant communication between the brain and body that makes this possible.

In horses, these messages travel through both nerves and hormones. The hypothalamus acts like a monitor, adjusting systems as conditions change. If a horse overheats, signals trigger sweating and shifts in circulation to cool the body. If stress rises, the stress response turns on; when safety returns, the body shifts back toward recovery.

behavioral flexibility and contributing to the kind of reactive, high-arousal behavior seen in Dante.

Changes to diet, turnout, and activity levels, like those Greg implemented, help restore the neuroendocrine rhythm and metabolic health. Movement facilitates **lymphatic flow**, glucose utilization, and the release of calming neuromodulators like serotonin and oxytocin. Fiber-based diets with low non-structural carbohydrates support gut-brain signaling and reduce glycemic spikes that can destabilize behavior. Vitamin B1, referenced here for its role in carbohydrate metabolism, may also modulate neural excitability and energy balance.

Dante's behavioral shift wasn't merely the result of groundwork or repetition; it emerged from neurobiological recalibration. By aligning management with the horse's evolved physiology, Greg allowed Dante's nervous system to return to balance (homeostasis), reducing metabolic stress, enhancing self-regulation, and restoring what looked like calm but was in fact biological stability.

LYMPHATIC FLOW

The lymphatic system is the body's "cleanup and defense" network. It carries fluid that helps remove waste, manage swelling, and support immune function. *Lymphatic flow* is simply the movement of this fluid through lymph vessels and lymph nodes. Unlike blood, lymph isn't pumped by the heart. It moves mainly through body movement and muscle activity. In horses, walking, grazing, and shifting weight help keep lymph flowing, while long periods of stall rest or inactivity can slow it down.

NEURO-INFLAMMATION

Neuroinflammation is the brain's immune response. In horses, special support cells (not typical white blood cells) detect trouble, clear debris, and release chemical signals. In the short term, this can be protective, helping the brain recover from injury, infection, or stress.

Problems arise when this response stays "on" too long. Chronic neuroinflammation can interfere with normal brain signaling and may show up as increased reactivity, reduced focus, or slower learning. Common drivers include ongoing stress, illness, pain, or poor environments.

CHAPTER 18

RUSTY

Surprise Rehab

MARK

In the spring of 2015, the movie script I had written sixteen years earlier, *Out of the Wild*, was finally green-lit. With financing in place, pre-production began immediately, and filming was scheduled to begin in September of that same year.

Out of the Wild is a modern day western, and we would be using our personal horses in the film to play the main equine roles. As we got farther along in pre-production, and the actors were being hired, it became clear that most of them had minimal horseback riding experience. The six horses we were bringing were all solid using horses, and I was confident that, if we could get the actors to show up a week or so before principal filming began, we could get them familiar with the horses and hopefully competent enough as riders so that they would look good in the saddle.

Working closely with the director during pre-production, I ended up doing some rewrites on the script that he suggested so we could stay within budget and finish filming on time. In doing so, it became clear that we were going to need at least one more "dead broke" horse that someone who barely knew how to hold the reins could get along with if need be. I began doing some horse shopping to see what was out there, with the plan being we would sell whatever horse we found after filming wrapped.

A few weeks later, I came across what appeared to be the perfect horse for what we would need. A thirteen-year-old former ranch gelding who had been ridden extensively on trails, been in parades, ridden by kids, and, most recently, was being used as a pony horse on the show circuit. That is, he ponied the show horses from the barn, paddock, or warm up pens to the show arena and back. His name was Rusty, due, I'm sure, to being a red roan. According to the owner, Rusty tied, could stand hobbled, was good for the vet and farrier, could cross water, good with his lateral work, great with a rope, had both leads, and could stop on a dime.

The videos that the seller sent seemingly confirmed everything he had told me about the gelding, and although Rusty was certainly compliant when he went through his paces, he didn't seem all that comfortable or happy. Still, the foundational pieces were there for what we needed, and I was confident we could at least help him feel a little more comfortable about his job leading up to filming.

We had Rusty shipped to our home in Colorado from Texas and were quite surprised at his overall condition when he arrived. He was dehydrated and about one hundred and fifty pounds underweight, he had saddle sores on either side of his withers with the one on his left side being much worse than the one on the right, had rain rot all over his back, and he was wearing worn out shoes on feet that looked to be at least four weeks overdue for trimming.

We also noticed almost right away that the point of his right hip, also known as the tuber coxae, the bony prominence on the upper, outer part of the hip (a landmark on the pelvis where muscles and ligaments attach) had been completely sheared off in some long-ago accident. The end of the broken bone had drifted downward and somehow reattached itself at a completely different angle about three inches lower than it was supposed to be. Luckily, this seemed to have little effect on his movement.

While all this suggested that Rusty's care at his former home may have been a bit on the unfortunate side, it was his overall demeanor that was a little more concerning. At first, Rusty didn't seem all that interested in food or water, and he certainly wasn't all that interested in interacting with people.

We put him in a stall with a run near where he could see our other horses, gave him plenty of hay and water, and left him alone overnight. The next morning, we found he had drunk most of the water in his twenty-gallon bucket and cleaned up all the hay we'd given him. He seemed much brighter but still not interested in people at all.

Over the next week or so, we introduced him to our herd by individually putting each horse in the stall and run next to his. Once the introductions were made and things calmed down between Rusty and one horse, we would introduce the next, and so on, until we'd introduced him to all six horses. Only then did we turn him in with them, making sure to keep an eye on their interactions.

The horses in our herd are pretty balanced, which was a good thing because Rusty was not. He charged into the herd like a dog chasing ducks, scattering them in every direction. After some squealing, and mostly superficial bites and kicks, things finally settled down. Our horses stayed together and gave Rusty plenty of space. This is how the seven of them would get along for the next few months.

Once permanently in the dry lot with our horses, another of Rusty's issues showed up: he wasn't interested in being caught. More specifically, it appeared he had been taught to stay away from humans when they approached. This became evident by the presence of what I would refer to as "trigger points" on his body. When it comes to catching, a trigger point is a particular spot or area on a horse's body which, when approached, will cause the horse to move away. These are very similar to trigger points on cattle.

For instance, when approaching Rusty from the side but moving slightly toward his head and at about twenty feet away, he would start turning his head away. The first trigger point was between his nose and cheek, and turning his head away exposed more of his neck and less of his face.

The second trigger point was near the base of his neck. When closing the distance to about eighteen feet, he would orient the base of his neck toward the person approaching. This orientation would cause him to take a step away, orienting the third trigger point toward the person. This was the one about in the middle of his rib cage and would show up about fifteen feet away. At thirteen feet away, he would present the fourth trigger point by moving his ribcage away and taking another step or two, exposing more of his flank and then hip toward the person. At ten feet, he would turn his entire body so his tail was toward the person (his tail being the fifth trigger point), and that was when he would leave, usually at a trot.

Trigger points like these get installed by (inadvertently) ignoring the signs that the horse is starting to get worried when we are, say, twenty feet away. If we miss the horse turning their head as a signal that they are concerned, we keep approaching. That causes them to turn away even more, until the pressure of us approaching becomes too much for them and they leave.

The more times we approach them while missing their signs of worry, the quicker they become at responding to the pressure they're feeling on those points. Eventually, we won't be able to get closer than twenty feet away before they leave. This is what was happening with Rusty.

We helped him resolve his catching issue by first and foremost acknowledging when he was very first starting to become worried. Then for a lack of a better term, we repurposed those trigger points. In other words, we helped him see that when someone reached one of those points, instead of worrying and turning away, he could relax and stay. Soon, he found even turning toward the person approaching wasn't such a bad thing.

The initial stages of this process, getting him to see he didn't have to leave whenever someone approached, took about an hour. We continued the reinforcement of that information for the rest of his life.

The next hurdle we came across with Rusty was his aversion to the farrier. Contrary to the assurances by his previous owner that he stood well to have his feet done, we found that not to be the case. This also caused me to wonder if that was the reason why he had worn-out shoes on feet that were way too long when he arrived.

The first time our farrier, Scott, got anywhere near Rusty, he immediately became worried to the point that attempting to work on him in that state of mind wasn't going to be in anyone's best interest. So even though he really needed to get his feet done, we decided not to do them that day, and instead, Crissi and I would work with his feet, and we would try again the next week when Scott would be back.

Either Crissi or I would spend time with Rusty's feet every day for the next week, and by the time Scott returned, Rusty stood nicely for him, even though he was still a little concerned. Not only is Scott a great farrier, but he is also an amazing horseman as well. He spent quite a bit of time as a trainer himself, and he worked for me as a ranch hand back during my ranch foreman days. As a result, he was able to feel his way through Rusty's troubled areas and gave him plenty of breaks so as not to get Rusty in a spot from which he might not be able to recover. Rusty never had a problem with his feet after that first time that Scott did them.

We were able to get Rusty's teeth balanced by our natural balance dentist, our friend Jim Masterson did a few bodywork sessions on him, we got a saddle that fit him, we put him on a balanced diet, and we even took him on the road with us during our summer clinic season. During that time, we were able to work on helping him soften up his responses to cues and clear up any misunderstandings he had about what was being asked of him under saddle.

In short, we treated and worked with Rusty the way we do all our horses. We try to help them feel better about things that are troublesome for them, and we build on the things that they do well.

Also, during that time, our herd slowly but surely accepted him as a member, and while he still had a little trouble from time to time, things were quieter than when he first arrived.

When September rolled around, we loaded up Rusty and the rest of our horses and hauled them to the production site in Nevada where *Out of the Wild* was to be filmed. Rusty did great with the inexperienced actors, basically babysitting them and helping them look pretty good on screen. We ended up using Rusty in several scenes, and when watching the film, the experienced eye would pick up on some of the overall concern Rusty was still carrying around.

The real turning point for Rusty and us came late one night a couple of weeks into filming.

This was the night we were going to film the opening sequence for the film. In it, our horse Rocky, who played the lead horse throughout the film, would be made up to look as though he was thin and injured and then be filmed from a distance to give the impression he was standing out in the desolate expanse of the desert.

The problem was, no matter how much makeup they used, Rocky was going to look too good for what the scene called for. He was in great shape, his coat was shiny, and he was muscled up so that he looked good for his other scenes.

The decision was made that we were going to need a stand in for Rocky, and the only other horse we had with us that looked remotely close to him was Rusty. They were the same height, roughly the same build and markings, and although Rusty was a red roan and Rocky was sorrel, their coloring looked almost indistinguishable in the dark. But the major selling point was that the desert heat had taken a bit of a toll on Rusty, and he'd dropped some weight during the shoot. His withers were prominent, and the old saddle sore on his left side could easily be made up to look like a serious injury.

The thing we weren't sure about was how Rusty would handle being put into that situation. After all, it wasn't all that long ago that he didn't want to be around humans at all and he was still leery about people he didn't know. On the set, he would be surrounded by people he didn't know, most of whom knew nothing about horses, while the crew went about the business of setting up the shot.

This entailed lights being moved around on cranes, cables being drug along the ground, the camera crew getting in place with their equipment, the makeup person applying "blood" to Rusty's face, neck, shoulder, withers, side, and hip, and all the while, Rusty was to stand quietly on his "mark."

The setup for the scene and getting Rusty's makeup on took nearly an hour, and with the camera crew about seventy-five yards away, it was finally time to get the shot. Rusty had been standing quietly in the same spot for that entire time, and I was well aware that we were in the middle of the desert and there was no telling how far it was to the nearest fence. I decided I'd go ahead and hobble Rusty while they filmed him, something he was comfortable with, and I'd keep his halter on until the very last minute. Then, to be on the safe side, I would lay on the ground a few feet away from him and hidden from the camera by the scrub brush, just in case he had any trouble and I needed to get to him.

When the time came, I slipped Rusty's halter off, gently stroked his forehead, then stepped away from him and laid down. The director yelled "action," and the desert went silent as the cameras rolled. About thirty seconds passed before he yelled "cut," but it seemed a lot longer.

Rusty hadn't moved a muscle. I got up and slipped his halter back on. I was taking off his hobbles so I could take him back to his buddies who were tied to the trailer about a quarter mile away when the director yelled, "Let's set up for the medium shot!" I wasn't aware we were doing any more filming that night, and with a medium shot being about half the distance between where the camera crew currently was and where Rusty and I were, that meant both Rusty's hobbles and me laying on the ground in front of him would definitely be in the shot. Obviously, that meant he'd have to stand even longer, this time without hobbles, and I wouldn't be able to be as close to him as I was for the last shot.

I stood with Rusty while the crew got ready, and, after about twenty minutes, they finally were. I slipped Rusty's halter off, stroked him on the forehead, and stepped slowly backward, stopping about ten feet away.

"You're in frame," the director told me. I backed away another ten feet.

"Still in frame," he said.

I backed up another ten feet. "Mark," It was Gareth, the camera operator this time. "Just

keep backing up until I let you know you're safe."

It was about here that Rusty looked as though he was going to start moving toward me. Noticing he was getting ready to move, I stepped toward him before he did, and he settled back. This was part of what we worked on months ago during our initial catching session. If he moved and I moved with him, my moving meant he could stop and relax.

As he stayed in place, I again started slowly backing away. Five feet, ten feet, twenty feet, thirty feet. "You're out!" Gareth said. I was now over sixty feet away from him and in no position to give him much help if he decided to leave in any direction other than toward me.

The director yelled, "Quiet on set." Then, "Action." Everything went quiet for another thirty seconds before the director finally yelled, "Cut!" I eased my way up to Rusty, who was exactly where I left him, slipped his halter back on, and whispered "Well done," to him while stroking him on his neck. Again,

I was getting ready to lead him away when the director told the crew to get ready for a close-up.

The crew took another twenty minutes, and while I would be able to stay relatively close to Rusty during this shot, there would be a new wrinkle. The camera would now be in the hands of the Steadicam operator. A Steadicam is a contraption strapped to the camera operator that allows the camera to move, float, and rotate during filming. For this shot, the operator got the camera within a foot of Rusty and then filmed the length of his body while "floating" the camera up and down next to him to film the "wounds" that the makeup guy had applied. At first, Rusty wasn't too sure about the Steadicam, which caused him to raise his head and turn slightly to look at it during the operator's dry run. But as soon as the worry showed up, I asked the operator to stop moving until Rusty settled again, which only took a few seconds.

The actual filming of the close-up went as smooth as the rest of the scenes that night, and I was finally able to halter Rusty and get him ready to go back to the trailer and his buddies. As I was buckling his halter, Crissi walked up. She was the script supervisor on the film and had been with the director the whole night.

"We can't sell him," she said, petting him on his forehead. "Not after all that." I agreed.

CHAPTER 19:

RUSTY

Returning to Safety

CRISSI

After almost a month of filming in the desert, I watched our friend Tim load four of our horses into his stock trailer. He would drop them off at our winter pasture in Colorado on his way home to New Hampshire. As I watched Rusty load, I remembered the day he arrived at our barn, only six months earlier.

When the shipper parked his rig and opened the stock trailer door, I saw a red roan horse who was underweight and looked exhausted. As if the driver had read my mind, he said, "I drove straight here from Texas. Sixteen hours with no breaks!" He was smiling, his ball cap stained with sweat. He held the door open while I untied and led Rusty out. I noticed the bucket hanging by his head was dry.

"When was the last time they were fed and watered?" I asked.

"Oh, somewhere in Kansas." He shut the door with a bang, got in his truck, and backed out, three more horses still in the trailer with what was probably a long day ahead of them.

Our first interactions with Rusty were making sure he ate and drank. A few days after that, I began leading him around the small property we lease before tying him to our trailer and brushing him. I added a beet pulp mash to his feed regimen. He was covered from withers to croup with rain rot, and the white hairs where there were saddle sores the size of my hand looked like the skin had been rubbed raw. I bathed him and applied some ointment.

Through all the daily handling, he was certainly quiet and compliant. But his eyes were dull, and he didn't protest about anything, once he was caught. He didn't express himself much, unless we had to work with his feet. After a couple of weeks, he'd gained weight, and the rain rot was slowly disappearing.

The first time I rode him was at a clinic in Utah. While Mark was working at one end of the arena, I saddled Rusty up and rode in the other. I wanted to get a feel for who he was and what he thought about being ridden. With every physical aid I applied, he was obedient. The odd quiet I felt from him was even more palpable while I was in the saddle, but it wasn't a quiet that felt peaceful. It seemed to me his quiet was as though he'd withdrawn into himself and his body was going through the motions.

I asked him into a trot to see what that was like. Again, he obeyed and jogged. When I asked him to come back to a walk, he kept jogging. And jogging. Head down, the rhythm never changed. It felt as though I was riding a robot. I picked up a rein and directed him into a smaller circle until he found his walk. His feet were too long to go faster than a jog. Another reason not to canter him was because it felt as though he was a horse a lot had been taken from. There are many horses like this, and the feeling that comes from them is one of being depleted. I dismounted.

Rusty felt as though he was being quiet because he had to be, not because he was comfortable.

Before filming started, we hauled Rusty to clinics, mostly so we could get to know him. He wasn't often ridden, instead either staying in a paddock with a pile of hay or tied to a hitch rail with a hay bag in front of him. Still, as Mark has noted, by the time that Rusty was filmed for the opening scenes of the movie, the desert heat had done a number on him, and he was once again thin. I was hopeful that once filming wrapped and we could get him out on pasture, he would gain weight before the cold weather set in.

Tim texted us when he dropped Rusty and four of our other horses off on pasture. There, they would be out on thirty-five grassy acres, with trees for shelter and a creek to drink from. Mark and I left Nevada and headed west, to a clinic in California. After we were done there, we drove for three days toward home, dropping our three clinic horses off at the same pasture. I remember standing beside Mark, watching as the herd took off at a gallop, Rusty rounder than I'd ever seen him. Mark and I agreed that the best thing for Rusty was to be left alone.

That spring, we went to get the herd off the pasture, Rusty stood beside Rocky, so I haltered them. I smiled when I saw how round and furry Rusty was. The angles of his hips were gone, his withers didn't look nearly as prominent, and he had a belly. What really made my day though, was the light in his eyes. He was still cautious when I touched his shoulder, leaning away from me, but he was also more present than I'd seen him six months earlier.

Another week went by while the horses settled back in at the barn. I brought Rusty out and tied him to the trailer. When I began grooming him, not only did he lean away, but he pinned his ear, his eye squinty and hard. I stayed where I was until he let some of the tension go, then I went and found softer brushes. He closed his eyes and sighed.

This was the first time that Rusty had expressed something to me. While I didn't want to encourage that tense state of mind, I also was inwardly overjoyed that he'd finally said something. That interchange was my opportunity to show him I was listening.

For years, Rusty taught me how to listen. I bought different saddle pads, wanting to be sure he was comfortable. His teeth were kept balanced, as were his feet. We made sure he always had food in front of him, received regular bodywork, and was fed joint supplements. I was careful to not ask too much. I spent as much time listening to him as I could, touching him more gently, buying him soft brushes, not putting myself too close to his head, as this seemed to cause him some nervousness.

As Rusty and I got to know each other, I could sense our partnership forming. I'd take him on walks through the forest when we were home. On clinic days, he'd spend as much time standing tied and eating as being ridden. I soon found out he had all kinds of skills...he just didn't know how to do them without physical or mental tension. We worked on relaxing into movement, and it wasn't long before he could sidepass, do a turn on the haunches, or do a walk to lope depart with intent and breath. He was easy to catch and easy to shoe, he loaded and unloaded well, and he settled into new places without much fuss. His behavior in the herd quieted down as well. Instead of going after them with hooves and teeth, he now pinned an ear or swished his tail.

Five years later during a hot August day, we were at Happy Dog Ranch. It was our weekend off, and we'd turned Rusty and two other of our horses out in a large grassy field. Several hours later, we went to catch them. They'd grazed their way to the far end. I called out, "Horses!" as I usually do when we went to see them on pasture. All three heads popped up out of the grass. Rusty took a couple of steps, rocketed into a gallop with ears forward and hooves pounding the earth, slid to a stop six feet in front of us, and then walked until he rested his head close to my chest, eyes closed.

One of my favorite movie lines comes from *Seabiscuit*. In it, the trainer is taking care of an old white horse. When asked why he didn't put the horse down, the trainer says, "You don't throw a whole life away just because it's banged up a little."

We're all a little banged up. Life can be messy and difficult for humans and certainly animals. When we got Rusty, it was only one year since my traumatic brain injury, and I was struggling with feeling confident with horses again. Rusty's body and his demeanor told us that he'd had a rough go in life. But with time and care, I learned that listening doesn't just happen with the ears or even with the eyes. Listening is what happens in our hearts too.

CLINICAL NOTES

DR. PETERS

Rusty's journey offers a compelling case study in neurobehavioral rehabilitation, and it also forces us to be precise about language. Some horses truly develop **learned helplessness**, a deeper state of resignation that follows repeated experiences of pressure, confusion, or discomfort without a workable escape. Other horses look similar on the surface but are better described as shutdown or temporarily offline from chronic stress, pain, depletion, or exhaustion. With Rusty, we cannot diagnose his full history from behavior alone, but the pattern fits a spectrum. His dull eyes, muted responsiveness, and "robotic" movement strongly suggested a hyper-aroused state with suppressed expression. Whether that suppression was long-standing learned helplessness or a more reversible shutdown, the clinical message is the same. Quiet is not always calm, and lack of protest is not proof of comfort.

> ## LEARNED HELPLESSNESS
>
> *Learned helplessness* can develop when a horse faces repeated pressure or stress it cannot escape or influence. Over time, the horse may stop trying, becoming unusually "quiet," dull, or compliant. This is not relaxation. It is resignation and loss of agency.

When Rusty arrived, the physical picture mattered. He was dehydrated, underweight, sore, and overdue in basic care. Those conditions alone can narrow a horse's behavioral range. A nervous system that is hurting, depleted, or chronically stressed often reduces exploration and communication because the body is prioritizing conservation. In horses, this can look like compliance, but internally, it may reflect autonomic dysregulation and neuroendocrine strain. The stress system can become overused, and recovery systems can become under-accessed. A horse may appear to "go along," not because he is okay, but because his system has learned that engagement does not improve outcomes.

Crissi's description in the saddle is especially revealing. Rusty obeyed every physical aid, yet the feel was not peaceful. He kept jogging when asked to walk, head down and with unchanged rhythm, as if he were running a program rather than participating. That kind of motor output often reflects a nervous system relying on patterned survival circuits with limited flexibility. It is consistent with hypoarousal and a freeze-like state, where the horse is present enough to comply but not available enough to modulate, explore, or offer nuance. Neurobiologically, these states are associated with altered limbic processing and shifts in the brain-body stress pathways, including circuits that involve the amygdala, the insula (which integrates internal body signals), and the hypothalamic stress axis. The key point for the reader is practical. A horse can look quiet while being internally dysregulated.

Rusty's difficulty being caught added another layer. Mark's "trigger points" illustrate how fear and avoidance become mapped in space. Rusty had learned, through repetition, that human approach predicts escalating pressure. The nervous system responded early and efficiently. He would turn the head, shift the neck, move the ribs, yield the hip, then leave. This is not stubbornness. It is conditioned avoidance shaped by the brain's threat circuits and spatial memory systems. The amygdala and hippocampus work together to link proximity with predicted outcomes, and the body organizes behavior to reduce risk.

The rehabilitation strategy worked because Mark and Crissi changed the prediction. Instead of pushing through those thresholds and forcing contact, they acknowledged the first signs of worry and worked just under Rusty's limit. Each time Rusty expected the approach to escalate and it did not, his nervous system received new information. Over repetitions, this is how fear associations soften. The trigger points were repurposed. Approach began to predict pause, relief, and safety rather than pressure. This is exposure done correctly. It stays inside the horse's processing window and gives the nervous system a way to succeed without flooding.

The farrier problem followed the same logic. Feet are vulnerable, restraint can feel trapping, and fear escalates quickly. The decision to pause, prepare Rusty daily, and return to the farrier when he could remain regulated protected both learning and safety. Scott's ability to give breaks and work within Rusty's recovery capacity prevented escalation and helped install a new association. Once Rusty experienced a successful foot-handling session that stayed within his threshold, the pattern did not need to return in the same way.

What ultimately changed Rusty was not one technique. It was a whole-body, whole-brain restoration. Hydration, nutrition, dental balance, hoof care, saddle fit, and bodywork reduced the internal drivers of guarding. Predictable routines reduced uncertainty. Time and low-arousal repetition gave the nervous system repeated evidence of safety. This is the biology of neuroplasticity. The brain reorganizes when experiences are consistent, manageable, and meaningfully different from what the horse previously predicted. Support cells and growth factors that underlie repair and learning are more likely to do their work when the basics are present. Adequate rest, lowered chronic stress, and physical comfort are not separate from training. They are the conditions that allow training to become learning.

Social buffering likely contributed as well. Rusty's herd initially gave him space, but over time, the steadiness of other regulated horses can help stabilize baseline arousal. As his physiology improved, his social behavior changed. He moved from charging and scattering the herd to using quieter, more normal signals, such as an ear pin or tail swish. That shift matters because it indicates increasing regulation and flexibility, not just "better manners."

One of the most important clinical markers in Rusty's recovery was the return of expression. When Crissi described him pinning an ear, leaning away, or squinting during grooming, she recognized something many people miss. In this context, the return of discomfort signals was not deterioration. It was a sign of reintegration. Rusty was becoming present enough to communicate. Crissi treated those signals as information and adjusted the experience, using softer brushes and gentler touch. That is how a horse learns that expression is safe, and that humans can be responsive rather than coercive. Over time, self-advocacy replaces suppression, and that is a meaningful welfare outcome.

The movie set scene became a powerful proof-of-change moment because it combined novelty, chaos, unfamiliar people, strange equipment, and long duration. Those conditions commonly reactivate old patterns. Rusty's ability to stand quietly for extended periods, tolerate makeup, tolerate equipment moving near him, and remain regulated off-halter reflected a nervous system with improved flexibility and recovery. Notice how the handling protected that state. When the Steadicam worried him, the movement stopped until he settled. When he began to follow Mark, Mark stepped toward him, a cue they had installed earlier that meant, "You can stop." The environment did not teach Rusty to be calm. The pattern of responsiveness, breaks, and predictability did.

Years later, Rusty running toward Crissi on pasture and resting his head near her chest captured the long arc of rehabilitation. At first, human approach predicted pressure, so he avoided. Later, human presence predicted regulation and safety, so he chose connection. That shift is not sentimentality. It is what nervous systems do when their predictions change. Rusty had not just learned new behaviors. He had changed how he felt about people, movement, and the world around him.

Rusty's story reminds us that the horse we see is never just muscle and movement. It is memory, wiring, internal state, and learned expectation. When we listen deeply and respond with patience, clarity, and respect for thresholds, we do not simply train behavior. We help rebuild a nervous system. And whether we label Rusty's early state as learned helplessness or a shutdown that left him temporarily offline, the clinical lesson holds. A quiet horse is not automatically a comfortable horse. The goal is not silence. The goal is a horse who is present, expressive, and able to recover into safety.

CHAPTER 20:

WILLOW

From Arenas to Acres

CRISSI

Every year we worked in Kansas, our clinic host Marilyn rode her reining mare Willow. A leggy chestnut with a flaxen mane and tail and four white socks, she was flashy at any gait.

The last year that we worked there, Marilyn invited us to dinner at her house and told us she was going to sell most of her herd due to a medical diagnosis she'd recently received.

"I mean, I'm not too old to ride," she laughed. "But the treatment I'm going to need will be long, and I'm not going to have a lot of energy to do much. I can't see how that's fair for my horses."

Mark and I listened, sympathizing with the choices she was forced to make.

"Can you take Willow?" Marilyn asked. "I know you'd give her a good home, and that means a lot to me. She means a lot to me."

Mark and I looked at each other, knowing that we'd recently been chatting about trying to find another clinic horse.

"It just so happens," Mark said, "that we have an extra spot in our trailer. We'd be happy to give Willow a home."

The day before we left, Marilyn wanted to show us what Willow could do. We watched as Willow was saddled, the lead rope looped through a tie ring bolted to the wall.

"She pulls back when I pull the girth tight," Marilyn said, in response to our question about tying.

Marilyn brought her saddle and pad out after grooming her mare, threw the saddle pad over, then lifted the saddle and dropped it on Willow's back. Even though the mare didn't move, she had her ears pinned, throwing in a few tail swishes. Her body was stiff, and I noticed she wasn't breathing very deeply.

We watched as Marilyn reached for the girth, threaded the latigo through the keeper twice, then pulled it as tight as she could get it. Willow arched her neck and stepped backwards before stopping and swishing her tail.

"Is that what she usually does?" Mark asked.

Marilyn said, "Yup. I just try to get the whole thing over with as quickly as possible." She patted the mare on her neck, then she took the bridle from a hook nearby and dragged the halter off her head, where it banged against the wall, then quickly put the bit in Willow's mouth, pushing her ears through the headstall and pulling her long blonde forelock through the browband.

By the time Marilyn took Willow to the arena and got on, we could see the mare was tense from nose to tail. Despite this, she willingly did whatever Marilyn asked, from walk-to-lope transitions to sliding stops, spins, and backing up. Marilyn rode over and said, "I've raised her from a two-year-old and started her myself." She patted the mare's neck. "She's fourteen now, but I never thought my last ride on her would be this soon."

When Mark and I brought Willow home, it was November, and our herd was already out on the thirty-five acre pasture they live on for six months of the year. There are trees for shelter and windbreak and two creeks on either end. We put a salt block out, and we check on them every week or two. We thought some downtime for Willow would be ideal, and since she was a good weight and had plenty of coat, we knew the herd would help her adjust.

What we didn't factor in was that besides her stall with a small run and the arena, Willow hadn't seen much of the world. She'd traveled around the state to reining competitions but hadn't been on trails or done anything besides reining patterns for at least ten years. When Mark and I turned her out on pasture, the other five horses watched then galloped up to meet her. We stayed for an hour, long enough to be sure she would be okay with them, then we drove home.

For the next couple of months, Willow seemed to adapt to her new life. But in January, we noticed that she'd lost enough weight that we could feel her ribs under her dull coat. We had a vet out, and the vet confirmed that Willow probably had sand in her gut and recommended a month of a psyllium supplement. We hauled Willow home that day and began the treatment.

Over the course of that month, she was visibly calmer, seeming relieved to be in a stall with a run. We added some extra nutritional support because it occurred to Mark and me that, besides the sand, she had maybe been stressed by the drastic change in surroundings. We decided to keep her at the barn until we brought the rest of the herd back in March.

While she was at the barn, we made sure she got her teeth balanced, had her hooves taken care of, got chiropractic and bodywork, and adjusted her nutrition to include gut support. Mark made sure his saddle fit her, working on putting on the pad and saddle quietly, the lead rope looped through the tie ring on the trailer. This took several days to sort out, but by the end, she could stand quietly. We hadn't tried tightening the girth, figuring we could work on that once we began working our clinics.

By the time we hauled Willow to begin our series of clinics at Happy Dog Ranch, we felt she was ready to be introduced to her new career as a clinic horse. The saddling went quietly, but as Mark drew up the girth and it touched her side, she arched her neck and pranced in place, sporadically blowing hard through her nostrils. Mark took the loose lead rope completely out of the tie loop, and waited for her to settle. Once she did, he loosened the already loose girth, and tried again, going slower. Once the girth was snug enough to keep his saddle on, he led her to the arena and taught while showing Willow what was expected of her: to follow quietly, stop when he stopped, and to relax instead of touching noses with whichever horse happened to be closest.

The good news was that Willow had no problems around other horses or being in an arena full of them, all moving at different speeds. Now we had something to build on.

We were, however, not surprised to find that, although she could stand quietly in an arena full of horses, she didn't know how to be led. It took more pressure to ask her to stop and an equal amount of pressure to ask her to follow us.

As the summer went on, Mark and I figured out how to help Willow with the girthing process, working on it not only at clinics, but several times a week at home. By August, Mark was riding Willow, figuring out that she didn't understand forward from leg pressure, which is common in horses who've been ridden with spurs. Without a leverage bit, Willow didn't understand how to stop, and she was certainly not relaxed while doing it. Turns were tense, and when asked to back up, Willow braced her whole body while her feet raised dust in the flurry to go backwards as quickly as possible.

During our last ten-day clinic in September, Willow understood how to walk with very little aid, stop in a relaxed state, turn by following her nose, and back up by letting go of tension first, and moving her feet second. I rode her a couple of times as well, noticing that moment when I could feel her understanding kick in. When that happened, we stopped, and she spent a couple of minutes not just licking and chewing, but yawning, closing her eyes, and stretching her head out. A couple of times she'd give a bone rattling shake while I was still on. It was a great practice for me to hone my understanding and perception of when we could ask a little more of whichever skill we'd like to practice, and when it was time to stop and let her sort it out.

By October, it was time to turn all the horses out again. This time, Willow's weight was good, and she'd grown a thicker coat than the previous winter. She'd integrated well into the herd, and our hope was that being away from a stall wouldn't be as stressful for her.

Each time we checked the herd, all of them were relaxed and hairy. Willow's body had changed from being drawn up and tight to relaxed and round. Part of this, we knew, was due to being out on grass and not being worked. Part of it, too, was that she felt calmer. We could tell this by the way she approached us with her head low and her barrel swinging, the wrinkles around her eyes and lips gone.

During the following clinic season, saddling and girthing Willow was much less stressful for her. Mark and I added walking and trotting in a calm state, as well as softening to the bit, and she latched on to this way of going almost immediately. We knew she could do some very fancy things, but what we also knew was that she'd been taught those behaviors (unintentionally) in a state of tension, both mental and physical. Our focus was making sure she could show us what she knew without the mental or physical tension attached. We ended her second year with us walking, trotting, stopping, and backing with relaxation. We added lateral work: turn on the forehand, turn on the haunches, and sidepass. By the end of the third year working with her, she could stand tied while being saddled and girthed, and we began working on helping her feel calm through the lope transition and into the lope itself.

Every year with Willow was our chance to show her that we'd do our best to be quiet, consistent, and clear. She followed our example by reflecting those lessons in everything she did. Even though she'd spent fourteen years with one person and knowing only a specific set of skills, by giving her time and showing her what her new job was, she revealed herself to be an exceptionally kind and confident mare.

Willow's transformation from a show horse to a clinic horse didn't happen overnight. It sometimes took days or weeks for her to trust our slower pace. Sometimes the external aids we added were confusing because they meant something else to her. But once she decided that we could be trusted, she became consistently quiet and levelheaded.

I have seen many horses being ridden a certain way—maybe too fast or too rough, maybe in a constant state of confusion. With some of their stories in this book, I hope we've shown that, given a life that supports their nervous system as well as their bodies, they all have the potential to return to their essential beingness.

Willow didn't owe it to us to cooperate; no horse does. It was up to Mark and me to consistently demonstrate that she was safe. She, in turn, gave us her trust. That, to me, is the art of horsemanship.

CLINICAL NOTES

DR. PETERS

Willow's transformation makes the most sense when we view her behavior through the brain systems that build routines. The cortex helps the horse notice, evaluate, and pay attention to what is happening. The striatum, deep in the brain, turns repeated experiences into automatic sequences. Together, these cortico-striatal loops are how horses learn patterns, predict what comes next, and carry out behavior quickly without having to "think it through" each time. In Willow's case, these loops explain how tension became part of everyday handling and how that tension was later replaced with a calmer program.

When Marilyn saddled Willow, the pinned ears, held breath, stiff neck, tail swishes, and stepping back were not personality flaws. They were a learned protective pattern. If the girth tightening repeatedly predicted discomfort, urgency, and lack of pause, Willow's nervous system would store that entire sequence as a routine. Once a routine is installed, it runs fast. The horse does not decide to brace. The body simply shifts into the pattern that previously helped her get through the moment.

Over time, patterns like this are often maintained by allostatic load, the cost of repeatedly adapting to stress without full recovery. The horse may still perform well, and Willow did. High-level reining maneuvers can be produced by deeply learned motor programs, even when the autonomic nervous system is operating in a state of tension. This is one of the central lessons of Willow's story. What a horse can do is not the same as how a horse feels while doing it.

When Willow was turned out onto thirty-five acres, her nervous system encountered a different kind of stress. Freedom is not automatically regulating if the environment is unfamiliar and difficult to predict. A horse who has lived in a stall-run-arena routine for years may have a nervous system built around a narrow world. Suddenly, there is open space, shifting herd movement, new footing, new smells, and constant novelty. That requires attention, scanning, and ongoing sensory processing. For some horses, the metabolic and emotional demand of that novelty shows up as weight loss, dull coat, vigilance, and unsettled social behavior. It is not that pasture is "bad." It is that the nervous system needs time and support to build new maps.

When Willow returned to the barn and seemed calmer, that makes biological sense too. Predictability lowers workload. When the world is legible again, the brain spends less energy on vigilance and more capacity becomes available for learning and recovery. Structure can be soothing when it is consistent and fair, especially for a horse carrying allostatic load.

Reworking Willow's saddling and girthing response required changing the prediction stored in her routine circuitry. Mark's approach did exactly that. He slowed the timeline, added pauses, softened the transitions, and waited for an exhale. He removed the sense of being trapped by taking the rope out of the tie loop when she became reactive, then reintroduced the girth gradually. This matters because high arousal pushes the horse into automatic defense and narrows learning. Lower arousal keeps the cortex online so the horse can register new information. Over repetitions, the girth no longer predicted "tighten and endure." It began to predict "pressure, pause, release." That is how old routines weaken and new ones form.

As Willow's foundation broadened, the same principle applied under saddle. She could do complex maneuvers, yet she did not understand softer basics like forward from leg, relaxed stopping without a leverage bit, or turning without tension. This is common when horses have learned sophisticated tasks inside a specific set of aids, often delivered in a state of intensity. Mark and Crissi rebuilt her system from the inside out. Walk, halt, turn, and back became opportunities to pair clarity with regulation. The goal was not simply to correct movement. The goal was movement that could occur while her nervous system stayed settled.

During those moments when understanding "clicked," Willow offered yawning, licking and chewing, stretching, and full-body shakes. These behaviors are often called calming signals, but their meaning depends on context. In training, licking and chewing commonly reflect a shift toward parasympathetic activity and the return of salivation after tension. Full-body shakes function as a neuromuscular reset, a way to discharge residual activation and reorganize posture. Yawning can appear during recovery as the system transitions, but yawning can also occur with discomfort, so it should always be interpreted alongside the whole horse and the situation. These behaviors are not sentimental messages. They are visible signs that the body is changing state.

By her second and third years, Willow showed what neuroplastic change looks like in real life. The saddling routine no longer triggered the old bracing loop. Her posture softened, her barrel swung, her eye and mouth lost the tightness that had become normal for her, and her physiology looked different across seasons. A thicker winter coat and more stable weight reflected improved regulation and recovery, not just calories. Her willingness to approach and her steadier presence reflected a nervous system that no longer expected conflict as the default.

Willow did not become a different horse. She became the same horse with a different set of predictions. When clarity is consistent, when pressure makes sense, when release is timely, and when the horse has room to recover, the brain updates. That is the deeper conclusion of her story, and it is a fitting final note for this book. Relationship does not replace training. Relationship is the biological context that allows training to become learning.

CHAPTER 21

THE FIVE DOMAINS

Putting it all together

DR. PETERS

For much of human history, "good horse care" was measured in the simplest terms: feed, water, shelter, and an absence of obvious injury. That is survival, and survival matters. Welfare science has clarified something essential: a horse can be well-managed and still not be well. Thriving is not an abstract ideal. It is a physiological and behavioral reality. A thriving horse is one whose daily life reliably supports comfort, safety, interest, social ease, and recovery, not merely the avoidance of harm. The Five Domains Model, developed by Professor David Mellor and colleagues, gives us a practical way to evaluate that fuller picture. Instead of asking only what suffering we prevented, it asks what experiences we made possible and what our management allows the horse to become.

From a neuroscience perspective, welfare is inseparable from prediction. The horse's nervous system is built to anticipate what comes next, continuously integrating internal signals, such as hunger, discomfort, and fatigue, with external signals, such as sound, movement, and social cues, and then comparing those inputs with memory. The central issue is not whether a horse experiences stress—stress is part of life—but whether the horse can repeatedly downshift into states that support digestion, restorative sleep, learning, and social connection. When the nervous system receives consistent evidence of safety and predictability, the horse can invest in recovery and exploration. When it receives repeated evidence of uncertainty, discomfort, or conflict, resources are diverted toward vigilance and defense. Over time, that allocation becomes a baseline, not a temporary mood, but the nervous system's default stance toward the world. This is why the Five Domains are best understood as an interacting system. Nutrition, Environment, Health, and Behavior are the daily conditions; Mental State is the integrated outcome because the brain is always converting conditions into felt experience. Mental State is not a vague add-on; it is the horse's internal report of how life is going.

Nutrition, in this model, is not simply "enough feed." It is one of the most direct ways we influence arousal, emotional steadiness, and behavioral stability because horses evolved as near-continuous foragers. Regular forage intake supports steady gut motility and a more stable internal chemistry. When forage is frequent, sufficient, and predictable, the nervous system receives a clear signal that resources are stable, and urgency decreases. When forage is restricted, delayed, or delivered in long gaps, the horse experiences repeated biological uncertainty, which often shows up as agitation, food guarding, reactivity, or preoccupation. These are not attitude problems; they are the behavioral footprints of an organism whose internal predictions are strained. Concentrates can be appropriate for certain horses and workloads, but high-sugar or high-starch patterns can challenge metabolic balance and, in vulnerable individuals, disrupt the gut-brain relationship in ways that amplify volatility. Thriving in this domain often looks like simple things done consistently: abundant forage, a steady rhythm, thoughtful spacing of resources, and a feeding strategy that reduces urgency rather than intensifying it.

The environment is the architecture of safety—the place where the nervous system lives in space and time. It includes shelter, footing, temperature, insects, airflow, and light–dark cycles, but it also includes what the brain cares about most: predictability, movement opportunity, and control over proximity. A horse is not merely housed in an environment; the horse is shaped by it. Space and movement are foundational inputs into regulation. When a horse can move freely, choose where to stand, and transition between rest and activity on its own terms, the brain receives steady evidence that the world is manageable. Environments that restrict movement, destabilize routines, or compress the horse into chronic proximity stress tend to keep the nervous system in watchfulness. A watchful brain may comply, but it settles less easily, sleeps more lightly, and learns with less flexibility. Good environments do not have to be perfect or expansive. They do need to be designed with the horse's sensory and locomotor biology in mind: space to move, places to rest, predictable routines, and features that reduce chronic background threat.

Health is the domain that reminds us pain changes the brain. Discomfort is not separate from behavior; it is among the most powerful drivers of behavior because it changes attention, lowers thresholds for reactivity, narrows emotional bandwidth, and makes learning more fragile. This is why training problems often improve when pain is addressed. The horse did not suddenly become cooperative; the horse became more physiologically capable. Because horses tend to conceal vulnerability, early signs are often subtle—small changes in posture, facial expression, stride symmetry, willingness, or transitions. When pain goes unrecognized, stress physiology can remain engaged, and inflammatory signals can interact with stress chemistry in ways that further erode emotional stability. Protecting welfare here means more than treating injuries. It means maintaining neural stability through timely assessment, good hoof and dental care, appropriate conditioning, and rehabilitation that rebuilds confidence as well as tissue.

Behavior is the life the horse is actually allowed to live: the expression of biology in motion. Horses are social, mobile grazers. When daily life blocks the patterns their nervous system expects—movement, foraging rhythm, social contact, rest in safety—stress often emerges as irritability, shutdown, hypervigilance, or repetitive coping behaviors. Social contact is not a luxury; stable companionship and predictable relationships support regulation, while isolation, inconsistent turnout, and chaotic mixing can destabilize baseline even in horses that seem manageable. This domain also includes human interaction because training is not merely instruction; it is a sequence of experiences that shapes the horse's predictions about people, pressure, and conflict. Low-conflict, consistent training strengthens trust and learning efficiency. Escalating, unpredictable, or coercive training strengthens defensive circuits and increases the likelihood of conflict behaviors. One of the most important welfare levers within behavior is agency, the horse's ability to make small choices and influence outcomes. Agency is not permissiveness. It is the biological opposite of helplessness, and it tends to shift the nervous system from bracing to participation. Horses with appropriate agency often become more curious and less defensive, not because they are "spoiled," but because their brains receive repeated evidence that the world is navigable.

Mental State is where the entire model becomes ethically meaningful because it asks the most honest question of all: how does the horse feel most of the time? A horse can look physically fine while living in chronic tension, anxiety, or resignation, states that quietly reshape sleep quality, muscle tone, attentional style, immune function, and the ease with which the horse can learn or relate. Conversely, when a horse repeatedly experiences comfort, safety, novelty within tolerable bounds, and stable social warmth, a different baseline emerges: more resilience, more interest, and greater availability for learning. Mental State is not a matter of opinion; it is the integrated consequence of the other domains. You can think of it as emotional weather, while nutrition, environment, health, and behavior are the climate systems that generate it. Improve the climate and, over time, the weather changes—not just for a day, but as a new default.

This is why systems like Jamie Jackson's "Pasture Paradise" have attracted attention in welfare conversations. Many modern horses live in static management: small paddocks, dry lots, stalls, limited turnout, or restricted social routines. Even with excellent intentions, these systems can press on every domain at once: less movement, disrupted foraging rhythm, fewer choices, fewer enriching experiences, and more time spent managing frustration. Track systems offer a practical alternative, especially on modest acreage. Instead of using one large square turnout as the entire living space, a track—often a loop along the perimeter—becomes the primary habitat, with resources placed strategically: water, hay stations, minerals, shade, scratching posts, and varied footing. The goal is not forced exercise; it is restoring a rhythm the horse's biology recognizes: move, graze, interact, rest, and repeat. Thoughtfully designed tracks can distribute forage and reduce binge cycles, add environmental variety without keeping the brain on edge, increase daily movement to support conditioning and gut motility, reduce conflict by dispersing resources, and promote calmer engagement through companionship and choice. From a brain perspective, enriched, choice-rich environments train the nervous system daily. Repeated safe exploration strengthens curiosity and problem-solving, predictable access to resources reduces threat loading, and social stability supports regulation. The track becomes more than a layout; it becomes a set of repeated experiences that shift baseline. And importantly, the concept scales: even small properties can create a meaningful loop, varied footing zones, and resource placement that invites movement and choice. The standard is not perfection; it is whether the design moves the horse closer to the life its nervous system expects.

The Five Domains, ultimately, remind us that welfare is not a checklist. It is a life. Every decision about food, space, pain, social access, and training becomes information the horse's nervous system uses to predict the world. Over time those predictions become the horse's baseline—calm or guarded, curious or resigned, socially open or defensive. When the horse's predictions are repeatedly met with safety, consistency, and appropriate freedom, the result is not just "better behavior." The horse becomes more physiologically able to rest, digest, connect, and learn. That internal change—quiet, cumulative, and measurable in the horse's way of being—is the foundation of ethical care, effective training, and the deeper partnership we all seek.

FINAL THOUGHTS:

THEIR WELFARE IS OUR PRIVILEGE

DR. STEPHEN PETERS

There comes a point when information no longer simply accumulates... it changes our perception. From that moment forward, what has been seen cannot be unseen.

The horse's brain is shaped by experience. Sensory input, social interaction, and environment contribute to how neural pathways are strengthened, weakened, or reorganized over time. This process never stops. Brains change in response to what is repeated, what is expected, what is feared, and what is allowed to be explored.

Change can occur through pressure and force, but that often results in rigidity, heightened threat sensitivity, and reduced capacity to adapt. Other forms of change emerge through curiosity, agency, choice, and social connection. These experiences expand the brain's ability to learn, recover, and remain flexible. In the most supportive conditions, the nervous system does not simply cope; it begins to thrive.

This understanding is why this book was written.

Equine welfare is not defined only by physical care, nutrition, or the absence of injury. It is defined by the internal state of the horse's nervous system and the quality of the world in which that system develops. A horse may appear calm on the outside yet live in a constant state of vigilance. Another may move freely, explore, connect, and rest, a reflection of a nervous system that feels safe enough to do so.

Welfare, therefore, becomes a question of environment and experience.

What does the horse's environment ask of its nervous system each day?

Does it support curiosity or suppress it?

Does it allow agency or remove it?

Does it provide social connection or isolation?

Going forward, the most important shift is not in method but in awareness. Every handler, rider, owner, veterinarian, and trainer has the capacity to help shape the sensory and social landscape of a horse's life. By creating

conditions that reduce chronic threat and support exploration, connection, and consistency, it becomes possible to support better behavior as well as better internal wellbeing.

This is where science becomes responsibility. Not the responsibility of perfection, but the responsibility of intention, informed by knowledge of how brains change.

The hope of this book is simple and profound: that it helps readers see the horse, not as a problem to be managed, but as a living nervous system in their care and that this vision will guide choices that allow more horses to live not in survival mode but in a state of safety, engagement, and possibility.

We don't owe horses perfection; we owe them a life their nervous system can learn to trust.

FINAL THOUGHTS:

A KINDER LIFE FOR HORSES

CRISSI MCDONALD

There have been pivotal moments in my lifetime with horses: happy moments, painful moments, unsure moments, excited moments, you name it; as horse people, we've faced them all.

What I'd like to share in closing are two instances, seeds we might say, for the book you hold in your hands.

When I attended a clinic that Mark was teaching, the first thing that caught my attention was his talking about lightness and softness as distinct and different states within the horse (and within us). Lightness is on the outside of the horse and only works if the horse isn't stressed. Softness, on the other hand, is carried on the inside of the horse and is available all the time. On the last day of the four-day clinic, Mark said "Softness is joy." Something inside me was shaken, like everything I'd learned about horses were scattering flakes in a shaken snow globe. Twenty-four years later, I continue to learn about the many facets of softness, both from people and horses.

The second moment was meeting Dr. Steve Peters. During that cozy presentation over dinner at his house, he said horses need two things: release and relief. I almost exploded out of my chair; here was that same snow globe feeling in my brain, shaken up, then rearranging.

When we educate ourselves, there are layers of advantages for others. In this case, for our friend, the horse. As you've now read, this doesn't mean only food, water, and shelter. There's robust science to support physical, mental, and emotional well-being through application of evidence-based methods. If spending my life with people and horses has taught me anything, it's that we all benefit when we're each supported and thus able to give our best.

I hope, after having read this book, that not only do you have answers to some of your own questions, but maybe new questions have been stirred up. That even though you've finished this book, you're not done exploring. That you'll look at your horses' lives and find ways to deepen your relationship with them, and honor who they are as beings.

FINAL THOUGHTS:

KNOW BETTER AND DO BETTER

MARK RASHID

The great western novelist Louis L'Amour told a story of how his young son was watching him while he was working on one of his books. The son asked him why he was typing so fast. Louis' response was, "Because I want to see what happens next."

That was exactly how I felt during the writing of this book. Our process consisted of either Crissi or me writing a story about a horse we'd worked with or owned, sending the story off to Steve, and then a few hours later, we would receive his clinical notes on the science behind what had been going on with the horse.

I have to admit, I found myself checking my email a little more regularly after sending Steve a story, in hopes those notes would have arrived. Without fail, the "ah-ha" moments would pile up as I found the information he would provide invaluable in understanding the underlying cause of behaviors each horse exhibited. I thought more than once how important that information was in eliminating the guesswork that can and often does accompany working with and being around horses.

I think it's important to understand that the horse world is full of training methods that are based around tradition, speculation, training "systems," educated guesses, made-up theories, and complete misunderstandings. Few are based around the horse's actual neurobiology, so inevitably many of these methods can and often do cause more problems than they solve.

I know for me personally, I can think of many horses over the years with whom I overstepped my bounds, believing what I was doing with them was correct based on what I knew at the time. I sure wish I had the information from this book back then.

We have to assume that folks who have always taught and practiced the kinds of methods I mentioned above do so because, like me, they truly believe they are doing the best by their horse. However, and as the old saying goes: "We should always do our best until we know better. Then we should do better."

I've come to understand that's what working on this book has done for me, and now that I know better, I can definitely do better. Having now spent time with *A Horse's Life,* I hope you feel you will be able to do better too.

ABOUT THE AUTHORS

DR. STEPHEN PETERS

Dr. Stephen Peters is a neuroscientist and board-certified clinical neuro-psychologist dedicated to bringing modern brain science into the world of horse training and care. He completed his practicum at Abbott Northwestern Hospital, an internship at Danbury Hospital, and a two-year postdoctoral fellowship at Hartford Hospital in the Departments of Neurorehabilitation, Neurology, and Neurosurgery. For more than a decade, he served as Chief of Neuropsychological Services in a major Neurology practice, where he contrib-uted to clinical research trials conducted under rigorous scientific protocols. He later founded and directed both The American Fork Hospital Memory Clinic and the Utah Valley Hospital Clinic for Brain Health.

Now retired from clinical practice, Dr. Peters unites decades of neuro-science expertise with a lifelong passion for horses. Through his Horse Brain Science Clinics offered worldwide, he provides classroom instruction, equine brain dissections, and live demonstrations that make complex neurobiology accessible and practical. His work empowers riders, trainers, veterinarians, and enthusiasts to understand equine behavior, learning, and welfare through the lens of evidence-based neuroscience.

A sought-after keynote speaker at scientific and medical conferences, Dr. Peters is the author of *Horse Brain Science* and *The Book of Neuropoetry*, and co-author of *Evidence-Based Horsemanship*. He lives at River Rock Ranch in Mancos, Colorado, with his wife, Michelle, their three Australian Shepherds, a cat, and six horses.

Learn more at **HorseBrainScience.info**.

MARK RASHID

Mark Rashid is an internationally recognized horseman and bestselling author known for his unique ability to understand the horse's perspective and resolve challenges through communication rather than force. His journey with horses began at age ten under the mentorship of an old cowboy who taught him to work *with* the horse, not against it, and to listen closely to what each animal is trying to say. That philosophy continues to shape Mark's work today. His clinics, built around individualized instruction for both horses and riders, attract participants from around the world.

In addition to his horsemanship, Mark has devoted decades to the study of the martial art of Aikido, earning a fourth-degree black belt. He teaches both at his local dojo and through Aibado workshops (Aikido for horseman), where he explores the parallels between softness, balance, and harmony in martial arts and working with students and their horses.

Mark spent many years working full-time on ranches gathering cattle, managing stock, and training horses, and he still enjoys helping on nearby ranches when time allows. He is the author of sixteen books, including *For the Love of the Horse* and numerous modern classics on horsemanship, as well as the novel *Out of the Wild,* which was adapted into a movie. Mark is also a musician with three recorded albums. When he's home, he enjoys building custom guitars and playing live music with local bands.

Learn more at **markrashid.com**.

CRISSI MCDONALD

Crissi McDonald is a clinician, author, photographer, and Masterson Method® Certified Equine Bodyworker whose work is rooted in a lifelong devotion to horses, storytelling, and visual art. For more than thirty years, these intertwined passions have shaped how she sees the world and how she helps others learn to see it differently too.

As an instructor and horse trainer with over four decades of experience, Crissi brings a broad and practical understanding of horse care, including dental maintenance, saddle fit, bits and bridles, nutrition, and how horses experience their environment. When a question stretches beyond her own expertise, she values collaboration and learning from trusted professionals who share her commitment to the horse's well-being.

Alongside her husband, renowned horseman Mark Rashid, she has traveled internationally for more than twenty years, working with thousands of horses and people across a wide range of disciplines. Her approach emphasizes softness, clarity of communication, and the deep learning that becomes possible in an atmosphere of safety and non-judgment.

Crissi's photography has been featured in numerous equine publications and books in the United States, the United Kingdom, and across Europe. As an author, she has written two nonfiction books exploring the horse-human bond, as well as the award-recognized romantasy series *North to Home*, which has earned multiple semifinalist and finalist placements in indie publishing contests.

Certified as a Masterson Method® Practitioner in 2017, Crissi has helped hundreds of horses find greater relaxation and ease. Whether through horses, words, or images, Crissi remains endlessly grateful for the chance to learn, share, and invite others into experiences that feel more at ease, both with their horses and themselves. She shares her life in Colorado with her husband Mark, four dogs, five horses, and a pony.

Learn more at **crissimcdonald.com**.

MORE BOOKS BY THESE AUTHORS

DR. STEPHEN PETERS

Evidence-Based Horsemanship (co-authored with Martin Black)
The Book of Neuropoetry
Horse Brain Science: The Neuroscience of Ethical Horsemanship

MARK RASHID

Considering the Horse
A Good Horse Is Never a Bad Color
Horses Never Lie
Life Lessons from a Ranch Horse
Horsemanship Through Life
Big Horses, Good Dogs, & Straight Fences
A Life with Horses: Spirit of the Work
Whole Heart, Whole Horse
Nature in Horsemanship: Discovering Harmony Through the Principles of Aikido
A Journey to Softness
Out of the Wild
Finding the Missed Path: The Art of Restarting Horses
For the Love of the Horse: Looking Back, Looking Forward
Something You Said
Richard the Chicken Eagle (children's book)

CRISSI MCDONALD

Continuing the Ride: Rebuilding Confidence from the Ground Up
Getting Along with Horses: An Evolution in Understanding
North to Home (fiction)
The Clock in the Water (fiction)
Rising from the Sea (fiction)

SELECTED REFERENCES AND SUPPORTING LITERATURE

Bachmann, I., Bernasconi, P., Herrmann, R., Weishaupt, M. A., & Stauffacher, M. (2003). Behavioural and physiological responses to an acute stressor in crib-biting and control horses. *Applied Animal Behaviour Science, 82*(4), 297–311.

Baribeau, D. A., & Anagnostou, E. (2015). Oxytocin and vasopressin: Linking pituitary neuropeptides and their receptors to social neurocircuits. *Frontiers in Neuroscience, 9,* 335.

Boissy, A., Manteuffel, G., Jensen, M. B., Moe, R. O., Spruijt, B., Keeling, L. J., Winckler, C., Forkman, B., Dimitrov, I., Langbein, J., Bakken, M., Veissier, I., & Aubert, A. (2007). Assessment of positive emotions in animals to improve their welfare. *Physiology & Behavior, 92*(3), 375–397.

Brubaker, L., & Udell, M. A. R. (2016). Cognition and learning in horses (*Equus caballus*): What we know and why we should ask more. *Behavioural Processes, 126,* 121–131.

Cabib, S., & Puglisi-Allegra, S. (2012). The mesoaccumbens dopamine in coping with stress. *Neuroscience & Biobehavioral Reviews, 36*(1), 79–89.

Corbetta, M., & Shulman, G. L. (2002). Control of goal-directed and stimulus-driven attention in the brain. *Nature Reviews Neuroscience, 3*(3), 201–215.

Craig, A. D. (2009). How do you feel? Interoception: The sense of the physiological condition of the body. *Nature Reviews Neuroscience, 10*(1), 59–70.

SELECTED REFERENCES

Crowell-Davis, S. L. (1993). Social behavior of the horse and its consequences for domestic management. *Equine Veterinary Education, 5*(3), 148–150.

Dias, B. G., & Ressler, K. J. (2014). Parental olfactory experience influences behavior and neural structure in subsequent generations. *Nature Neuroscience, 17*(1), 89–96.

Draaisma, R. (2020). *Language signs & calming signals of horses.* CRC Press.

Eagleman, D. (2015). *The brain: The story of you.* Vintage Books.

Feinstein, J. S., Adolphs, R., Damasio, A., & Tranel, D. (2011). The human amygdala and the induction and experience of fear. *Current Biology, 21*(1), 34–38.

Fiedler, D., Reiser, M., Weitlauf, C., & Ratté, S. (2021). Brain-derived neurotrophic factor/tropomyosin receptor kinase B signaling controls excitability and long-term depression in the oval nucleus of the BNST. *Journal of Neuroscience, 41*(3), 435–445.

Flagel, S. B., Clark, J. J., Robinson, T. E., Mayo, L., Czuj, A., Willuhn, I., Akers, C. A., Clinton, S. M., Phillips, P. E. M., & Akil, H. (2011). A selective role for dopamine in reward learning. *Nature, 469,* 53–57.

Gallo, E. F., Meszaros, J., Sherman, J. D., Chohan, M. O., Teboul, E., Chuhma, N., & Kellendonk, C. (2018). Accumbens dopamine D2 receptors increase motivation by decreasing inhibitory transmission to the ventral pallidum. *Nature Communications, 9,* 1086.

Gleerup, K. B., Forkman, B., Lindegaard, C., & Andersen, P. H. (2015). An equine pain face. *Veterinary Anaesthesia and Analgesia, 42*(1), 103–114.

Grahn, J. A., Parkinson, J. A., & Owen, A. M. (2009). The role of the basal ganglia in learning and memory: Neuropsychological studies. *Behavioural Brain Research, 199*(1), 53–60.

Graybiel, A. M. (2008). Habits, rituals, and the evaluative brain. *Annual Review of Neuroscience, 31,* 359–387.

Horvitz, J. C. (2009). Stimulus–response and response–outcome learning mechanisms in the striatum. *Behavioural Brain Research, 199*(1), 129–140.

Insel, T. R. (2010). The challenge of translation in social neuroscience: A review of oxytocin, vasopressin and affiliative behavior. *Neuron, 65*(6), 768–779.

Keeling, L., Jonare, L., & Lanneborn, L. (2009). Investigating horse–human interactions: The effect of a nervous human. *The Veterinary Journal, 181*(1), 70–71.

Leblanc, M. (2013). *The mind of the horse: An introduction to equine cognition.* Harvard University Press.

McGreevy, P. (2004). *Equine behavior: A guide for veterinarians and equine scientists.* Saunders.

Peters, S. (2025). *Horse brain science: The neuroscience of ethical horsemanship.* Wasteland Press.

Porges, S. W. (2011). *The polyvagal theory: Neurophysiological foundations of emotions, attachment, communication, and self-regulation.* W. W. Norton & Company.

Sapolsky, R. M. (2004). *Why zebras don't get ulcers* (Rev. ed.). St. Martin's Press.

Schultz, W. (2016). Dopamine reward prediction error signaling: A two-component response. *Nature Reviews Neuroscience, 17*(3), 183–195.

Siegel, D. J. (1999). *The developing mind: How relationships and the brain interact to shape who we are.* The Guilford Press.

Zhang, X., Liu, Y., Li, J., Gao, C., Wang, Y., & Wang, J. (2025). Dopamine induces fear extinction by activating reward-responding amygdala neurons. *Proceedings of the National Academy of Sciences, 122*(18), e2319173121.

GLOSSARY

Amygdala: helps detect biologically important events, especially possible threat, and assign emotional significance.

Basal Ganglia: a group of deep brain structures involved in action selection, movement, habit formation, motivation, and learning.

Brain-Derived Neurotrophic Factor (BDNF): a protein that supports the growth and strengthening of neural connections.

Caudate Nucleus: part of the basal ganglia's cognitive loop, involved in attention, action selection, and goal-directed behavior.

Cerebellum: helps coordinate movement, timing, learning, and some aspects of cognitive and emotional regulation.

Cingulate Cortex: involved in attention, emotion, motivation, and decision-making.

Conditioned Automatic Response: a learned response that becomes fast and automatic through repetition.

Fear Extinction: learning that a cue or situation no longer predicts danger, forming a new safety memory.

Freeze: a defensive state of stillness under threat, often with heightened vigilance, heart-rate slowing, and relative parasympathetic influence; it is not the same as shutdown or learned helplessness.

Frontal Cortex: the front part of the brain involved in attention, inhibition, flexibility, and adjusting behavior.

Gamma-Aminobutyric Acid (GABA): the brain's main inhibitory neurotransmitter, helping reduce neural activity.

Habituation: reduced response after repeated exposure to a neutral, non-threatening stimulus.

Hippocampus: involved in memory, context, and spatial mapping.

Homeostatic Signaling: ongoing brain-body communication that helps maintain internal balance.

Hypothalamus: helps regulate hunger, thirst, temperature, hormones, and autonomic balance.

Inbreeding: breeding between closely related animals, increasing the chance of expressing harmful recessive traits.

Insula: a brain region involved in sensing and integrating internal body states, contributing to feeling, salience, and self-regulation.

Learned Helplessness: a more enduring pattern that can develop after repeated, uncontrollable aversive experience, in which the animal cannot escape despite its attempts and eventually stops trying to influence outcomes.

Limbic System: a general term for interconnected brain regions involved in emotion, memory, motivation, and learning.

Long-Term Potentiation (LTP): a lasting strengthening of connections between neurons that supports learning and memory.

Lymphatic Flow: the movement of lymph through vessels and nodes, helping regulate fluid balance, remove waste, and support immune function.

Myelination: formation of myelin around nerve fibers, helping signals travel more efficiently.

Neuroinflammation: activation of the brain's immune response, often involving glial cells and inflammatory signaling.

Neuropeptides: small signaling molecules that often act more slowly and last longer than classical neurotransmitters.

Neuroplasticity: the nervous system's ability to change with experience.

Norepinephrine: a neuromodulator involved in arousal, vigilance, attention, and readiness for action.

Oxytocin: a neuropeptide involved in social bonding, maternal behavior, and aspects of stress regulation.

Parasympathetic Nervous System (PNS): the branch of the autonomic nervous system associated with rest, digestion, energy conservation, and recovery.

Place Cells: hippocampal neurons that become active in relation to a particular location.

Positive Prediction Error: a learning signal generated when an outcome is better than expected.

Procedural Learning: learning skills and routines through repeated practice until they become more automatic.

Putamen: part of the basal ganglia's motor loop, involved in movement patterns, motor learning, and habitual action selection.

REM Sleep (Rapid Eye Movement): a stage of sleep associated with active brain processing, memory integration, and emotional processing.

Reward-Mediated Fear Extinction: extinction learning in which reward helps strengthen a new, safer association.

Sensory Integration: the process by which the nervous system combines sensory and bodily information to guide perception and action.

Serotonin: a neurotransmitter involved in emotional balance, impulse regulation, and recovery from stress.

Shutdown: an outward state in which a horse appears dulled, withdrawn, or minimally responsive because the nervous system is overwhelmed; it may be temporary and should not be mistaken for relaxation or understanding.

Slow-Wave Sleep (SWS): deep non-REM sleep important for restoration and memory consolidation.

Social Buffering: reduction of stress responses in the presence of a familiar, bonded, or calm companion.

Ventral Striatum: a brain region involved in motivation, reward learning, and linking expected outcomes to action.

Vomeronasal Organ (VNO): an accessory chemosensory organ involved in detecting certain social and reproductive chemical cues.